微服务开发实战

[美] 保罗·奥斯曼 著
邓 彪 译

清华大学出版社
北 京

内 容 简 介

本书详细阐述了与微服务相关的基本解决方案，主要包括单体架构应用程序分解、边缘服务、服务间通信、客户端模式、可靠性模式、安全性、监控和可观察性、扩展、部署微服务等内容。此外，本书还提供了相应的示例、代码，以帮助读者进一步理解相关方案的实现过程。

本书适合作为高等院校计算机及相关专业的教材和教学参考书，也可作为相关开发人员的自学教材和参考手册。

北京市版权局著作权合同登记号 图字：01-2019-1825

Copyright © Packt Publishing 2018.First published in the English language under the title
Microservice Development Cookbook.
Simplified Chinese-language edition © 2021 by Tsinghua University Press.All rights reserved.

本书中文简体字版由 Packt Publishing 授权清华大学出版社独家出版。未经出版者书面许可，不得以任何方式复制或抄袭本书内容。

本书封面贴有清华大学出版社防伪标签，无标签者不得销售。
版权所有，侵权必究。举报：010-62782989，beiqinquan@tup.tsinghua.edu.cn。

图书在版编目（CIP）数据

微服务开发实战 /（美）保罗·奥斯曼（Paul Osman）著；邓彪译. —北京：清华大学出版社，2021.6
书名原文：Microservice Development Cookbook
ISBN 978-7-302-58185-7

Ⅰ. ①微… Ⅱ. ①保… ②邓… Ⅲ. ①互联网络—网络服务器 Ⅳ. ①TP368.5

中国版本图书馆 CIP 数据核字（2021）第 094611 号

责任编辑：贾小红
封面设计：刘　超
版式设计：文森时代
责任校对：马军令
责任印制：朱雨萌

出版发行：清华大学出版社
网　　址：http://www.tup.com.cn，http://www.wqbook.com
地　　址：北京清华大学学研大厦 A 座　　邮　编：100084
社 总 机：010-62770175　　邮　购：010-62786544
投稿与读者服务：010-62776969，c-service@tup.tsinghua.edu.cn
质量反馈：010-62772015，zhiliang@tup.tsinghua.edu.cn

印 装 者：三河市吉祥印务有限公司
经　　销：全国新华书店
开　　本：185mm×230mm　　印　张：15.5　　字　数：311 千字
版　　次：2021 年 7 月第 1 版　　印　次：2021 年 7 月第 1 次印刷
定　　价：89.00 元

产品编号：080241-01

译 者 序

现代软件开发越来越多地强调持续集成（CI）和持续交付（CD），这和当前的市场需求及技术发展是离不开的。就像现代工厂里的装配线以快速、自动化和可重复的方式从原材料生产出消费品一样，软件交付管道也需要以快速、自动化和可重复的方式从源代码生成可发布版本。该模式被称为"敏捷开发"，启动装配线的过程被称为"持续集成"，而完成这项工作的总体设计被称为"持续交付"，确保质量的过程被称为"持续测试"，而将最终产品提供给用户的过程则被称为"持续部署"。

以游戏开发为例，早期的游戏采用的是单体架构模式，开发测试流程特别长，一款游戏开发经历3～5年是常有的事，就连更新一个资料片也可能以年为单位计时，甚至一再跳票。而在页游和手游时代，一款游戏可能仅需要一个月就可以交付测试版，开放给玩家进行测试，然后再根据玩家的游戏行为和反馈意见增删功能，快速迭代。而要实现这种敏捷开发，最好的方式就是采用微服务架构。

本书是非常实用的微服务开发技术指南。书中首先介绍如何将单体架构应用分解为微服务、迁移生产环境中的数据、将单体架构应用程序升级为服务，从而为转换到微服务架构打下基础；然后以一个图像消息程序的开发为例，介绍各种工具及其实际编码操作。例如，使用嵌入式Zuul服务来提供边缘代理功能、使用Netflix边车项目和Eureka扩展服务、使用Spring Boot和Hystrix在Java中构建一个小型示例网关服务、使用Hystrix停止级联故障、配置NGINX以进行速率限制、安装和配置Linkerd以控制Ruby on Rails单体架构API与服务之间的通信、使用ZooKeeper构建服务发现机制、使用AWS CLI工具创建自动缩放组并附加弹性负载均衡解决方案、使用Ribbon实现客户端负载均衡、使用Kafka构建事件驱动的微服务、使用CompletableFuture以及Java 8流的组合来对依赖性的服务调用进行建模、设计和构建服务于前端的后端（BFF）层、使用Thrift或gRPC实现一致的远程过程调用（RPC）、使用resilience4j库的断路器实现提高可靠性、使用指数退避算法避免惊群效应、使用Redis建立缓存以提升性能、使用Fastly内容分发网络、通过游戏日演习验证容错能力（附游戏日演习模板）、使用Chaos toolkit引入自动化混沌工程、使用自定义seccomp策略运行容器、使用Terraform开源工具提供AWS资源的集合、使用StatsD和Graphite收集度量指标、使用Prometheus开源监控和警报工具包、在Docker容器中运行Jaeger分布式跟踪系统、配置Alertmanager发送警报、使用Vegeta和Gatling对微服务进行负载测试、使用 AWS 创建自动扩展计算集群、使用 Docker Compose 运行多容器应用程序、在 Kubernetes 上部署服务和使用金丝雀部署方式测试版本等。

在翻译本书的过程中，为了更好地帮助读者理解和学习，本书以中英文对照的形式保留了大量的术语，这样的安排不但方便读者理解书中的代码，而且也有助于读者通过网络查找和利用相关资源。

本书由邓彪翻译，陈凯、唐盛、马宏华、黄刚、郝艳杰、黄永强、黄进青、熊爱华等也参与了本书的部分翻译工作。由于译者水平有限，错漏之处在所难免，在此诚挚欢迎读者提出意见和建议。

译　者

前　　言

关于微服务

在过去几年中，微服务已成为越来越受欢迎的主题。与任何新的架构概念一样，人们对微服务也有很大的误解，甚至微服务（Microservices）一词本身也令人困惑，新手通常难以分辨清楚微服务的适当大小（提示：实际上它并不是指代码库的大小），并且可能会陷入如何开始使用这种架构风格的困境。

面向服务的架构并不是什么新鲜事物。20 世纪 90 年代，各种公司都在推广 Web 服务，以解决越来越大型化的、僵化的代码库问题。Web 服务承诺将提供可重用的功能，代码库可以轻松使用这些功能。诸如 SOAP 和 WSDL 之类的技术开始获得采用，但它们似乎从未实现易用性承诺。与此同时，诸如 PHP、Ruby 和 Python 之类的开源语言以及诸如 Symfony、Rails 和 Django 之类的框架使得开发单体架构的以 Web 为中心的代码库变得更加容易。

就这样过了几十年，我们又重新对服务产生了兴趣。那么，为什么会有这种变化？首先，随着富 Web 和移动应用程序的出现，每个系统当前都是分布式系统。由于云计算的出现，计算和存储资源比以往任何时候都便宜。容器正在改变我们对部署和运营服务的思考方式。许多消费者服务已经超出了其单体架构代码库的规模，并且团队也发现单体架构应用程序难以扩展。

微服务的出现，可以帮助解决许多问题。

采用微服务架构的先决条件

微服务并不是万能的。尽管它们有很多好处，但是它们也带来了一些特定的挑战。在决定转向微服务之前，重要的一点是要拥有一定的基础设施和工具。Martin Fowler 和 Phil Calcado 都撰写过有关微服务先决条件的文章。Martin Fowler 的文章网址如下：

https://martinfowler.com/bliki/MicroservicePrerequisites.html

Phil Calcado 的文章网址如下：

http://philcalcado.com/2017/06/11/calcados_microservices_prerequisites.html

在此，我们不想鹦鹉学舌重复别人讲过的东西；相反，我只想强调，在开始开发微服务之前，必须进行一定程度的自动化和监控。你的团队应该乐于承担值守职责，并且你应该拥有一个用于管理警报和升级的系统，如 PagerDuty（http://pagerduty.com/）。

微服务的优势

采用微服务架构可以获得多种好处，以下我们将详细讨论。

伸缩性

在单体架构代码库中，不容易实现伸缩性，导致铺张过大、浪费资源，或者捉襟见肘、不敷使用。而微服务则不存在这个问题，它可以轻松地根据自己的需求扩展应用程序的各个部分。例如，应用程序的某个特定部分可能是每个用户都会用到的（如身份验证/授权服务），而另一部分则仅由部分用户使用（如搜索或发送消息）。不同的流量模式将转化为不同的扩展需求以及应用于扩展服务的不同技术。我们可以针对用户的每个请求开发服务，使得数据读取和存储的成本更低。另外，对于不需要提供非常一致结果的服务还可以利用缓存技术，这样响应速度更加快捷。

团队组织

当工程师团队在采用单独部署的单独代码库上工作时，他们能够独立地做出很多决定，而无须与组织中的其他团队进行协调。这意味着工程师可以自由地提交代码，设计自己的代码审查流程并部署到生产中，而无须始终进行协调。在单体架构应用程序中，工程师不得不将其更改放入队列中，然后在特定时间与其他团队的更改一起部署，这种情况非常普遍。如果出现问题（有害部署是最常见的中断原因之一），则整个变更集将被回滚，从而延误了多个团队的工作。微服务允许团队以更大的自主权来帮助你避免这种情况。

可靠性

当单体架构应用程序失效时，它往往会完全失效。例如，首先是数据库不可用，然后该应用程序会尝试在连接池中使用陈旧的连接，最终，服务请求的线程或进程被锁定，并且使用户白屏死机或无法运行移动应用程序。微服务使你可以根据情况决定如何处理应用程序特定部分的故障。如果你的服务无法访问数据库,则最好返回一个过时的缓存或一个空响应;如果你的服务已经失败并开始返回 HTTP 503 响应,则上游服务可以通过施加背压来响应,从而使服务能够被恢复。

微服务为你提供了更大的自由来隔离系统中的故障,从而为用户带来更愉快的体验。

本书将为你开发微服务时可能涉及的许多主题提供方便的参考。我们将以秘笈的方式呈现开发技巧,这些秘笈将帮助你从单体架构顺利转换到微服务架构。我们将讨论在选择最佳架构和管理微服务方式时遇到的特定问题或面临的挑战。本书秘笈包含大量有效、简单、经过测试的示例,你可以将它们应用到自己的开发过程中。我们希望本书可以帮助你思考、计划和执行基于微服务的应用程序的开发。祝阅读愉快!

本书读者

如果你是想构建有效且可扩展的微服务的开发人员,那么本书非常适合你。本书假定你具有微服务架构的一些基础知识。

内容介绍

本书共包含 9 章,具体内容如下。

第 1 章"单体架构应用程序分解",将详细介绍如何实现从单体架构应用程序到微服务的过渡,本章秘笈着重于架构设计。当你开始使用微服务这种新的架构风格开发功能时,你将需要了解如何应对一些最初的挑战。

第 2 章"边缘服务",教你如何使用开源软件将服务公开到公共互联网、控制路由、扩展服务的功能,以及在部署和扩展微服务时如何应对许多常见挑战。

第 3 章"服务间通信",将详细讨论使你能够自信地处理微服务架构中必定需要的各种交互的方法。

第 4 章"客户端模式",将讨论对依赖的服务调用进行建模并聚合来自各种服务的响应以创建特定于客户端的 API 的技术。此外,我们还将讨论如何管理不同的微服务环境,使用 JSON 和 HTTP 实现 RPC 一致性,以及 gRPC 和 Thrift 的使用。

第 5 章 "可靠性模式"，将讨论许多有用的可靠性模式，这些可靠性模式可以在设计和构建微服务时使用，以减少预期和意外的系统故障的影响。

第 6 章 "安全性"，包括身份验证微服务、确保容器安全等秘笈，可帮助你学习构建、部署和操作微服务时要考虑的许多最佳实践。

第 7 章 "监控和可观察性"，将详细介绍若干个监控和可观察性方面的功能。我们将演示如何修改服务以发出结构化日志。此外，还将研究度量指标，使用许多不同的系统来收集、汇总和可视化度量指标。

第 8 章 "扩展"，将讨论使用不同工具的负载测试。我们还将在 AWS 中设置自动伸缩组，使它们可以按需扩展。

第 9 章 "部署微服务"，讨论了容器、业务流程和调度，以及将更改安全地交付给用户的各种方法。本章中的秘笈应作为一个良好的起点，特别是如果你习惯于在虚拟机或裸机服务器上部署单体架构应用程序的话。

充分利用本书

阅读本书，假定你具有一些有关微服务架构的基础知识。另外，本书还需要安装若干软件包，在相应的操作秘笈中都提供了这些安装操作的说明或网址。

下载示例代码文件

读者可以从 www.packtpub.com 下载本书的示例代码文件。具体步骤如下。

（1）登录或注册 www.packtpub.com。

（2）在 Search（搜索）框中输入本书名称 *Microservices Development Cookbook* 的一部分（不分区大小写，并且不必输入完全），即可看到本书出现在推荐下拉菜单中，如图 P-1 所示。

（3）单击选择 Microservices Development Cookbook 这本书，在其详细信息页面中单击 Download code from GitHub（从 GitHub 中下载代码）按钮，如图 P-2 所示。需要说明的是，你需要登录此网站才能看到该下载按钮（注册账号是免费的）。

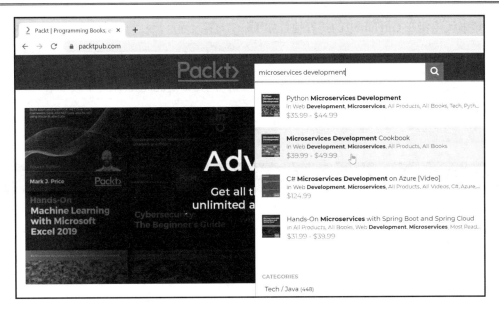

图 P-1

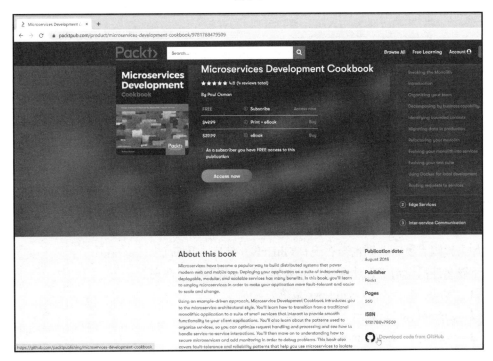

图 P-2

该书代码包在 GitHub 中的托管地址如下：

https://github.com/packtpublishing/microservices-development-cookbook

在该页面上单击 Code（代码）按钮，然后选择 Download ZIP 即可下载本书代码包，如图 P-3 所示。

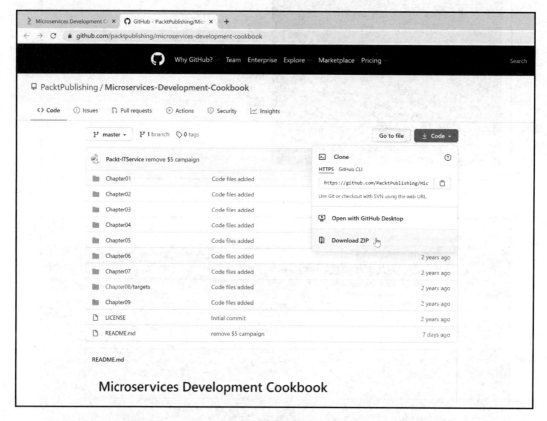

图 P-3

如果代码有更新，则也会在现有 GitHub 存储库上更新。

下载文件后，请确保使用最新版本解压缩或解压缩文件夹。

- ❏ WinRAR/7-Zip（Windows 系统）。
- ❏ Zipeg/iZip/UnRarX（Mac 系统）。
- ❏ 7-Zip/PeaZip（Linux 系统）。

本书约定

本书使用了许多文本约定。

（1）CodeInText：表示文本中的代码字、数据库表名、文件夹名、文件名、文件扩展名、路径名、虚拟 URL 和用户输入等。以下段落就是一个示例：

> 无论是处理微服务架构还是单体架构代码库，都需要从根本上解决系统在某种故障状态下正常运行的问题。可以访问以下链接以获取更多信息：
>
> https://www.youtube.com/watch?v=tZ2wj2pxO6Q

（2）有关代码块的设置如下：

```
class AttachmentsService

    def upload(message_id, user_id, file_name, data, media_type)
        message = Message.find_by!(message_id, user_id: user_id)
        file = StorageBucket.files.create(
            key:    file_name,
            body:   StringIO.new(Base64.decode64(data), 'rb'),
            public: true
        )
```

（3）当要强调代码块的特定部分时，相关行或项目以粗体显示。示例如下：

```
dependencies {
    testCompile group: 'junit', name: 'junit', version: '4.12'
    compile group: 'io.github.resilience4j', name: 'resilience4j-circuitbreaker', version: '0.11.0'
    compile group: 'org.springframework.boot', name: 'spring-boot-starter-web'
}
```

（4）任何命令行输入或输出都采用如下所示的粗体代码形式：

brew install docker-compose

（5）术语或重要单词采用中英文对照形式，在括号内保留其英文原文。示例如下：

> 康威定律（Conway's Law）告诉我们，组织设计的系统架构是组织本身的沟通结构的副本，也就是说，产品的系统架构将打上开发它的组织的沟通结构的烙印。这通常意味着，软件工程团队的组织结构将对其开发的软件的设计结构产生深远的影响。

（6）本书还使用了以下两个图标。

🅘表示警告或重要的注意事项。

💡表示提示或小技巧。

编写体例

本书大多数章节以秘笈形式编写，每一节就是一个秘笈，每个秘笈中又分别包括"理论阐释""做好准备""实战操作"等小节，使你既能理解微服务开发相关知识和原理，又能带着问题进入实战编码阶段，实现有目的且高效的学习。

关于作者

保罗·奥斯曼有十多年的构建内部和外部平台的经验。从面向第三方的公共 API 到内部平台团队，他帮助构建了支持大型消费者应用的分布式系统。他拥有管理多个工程师团队的经历，致力于快速交付基于服务的软件系统。

保罗·奥斯曼发表过多篇有关微服务和运维一体化的技术文章，并进行了多次主题演讲。他是开放技术平台和工具的热情拥护者。

目　　录

第1章　单体架构应用程序分解 1
1.1　导语 1
1.2　组织开发团队 1
1.2.1　实战操作 2
1.2.2　示例讨论 3
1.3　按业务功能分解微服务 3
1.3.1　理论阐释 4
1.3.2　实战操作 4
1.4　识别有界上下文 5
1.4.1　理论阐释 5
1.4.2　实战操作 8
1.5　迁移生产环境中的数据 9
1.5.1　理论阐释 9
1.5.2　实战操作 9
1.6　重构单体架构应用程序 12
1.6.1　理论阐释 12
1.6.2　实战操作 14
1.7　将单体架构应用程序升级为服务 16
1.7.1　理论阐释 16
1.7.2　实战操作 16
1.8　升级测试套件 18
1.8.1　做好准备 18
1.8.2　实战操作 18
1.9　使用Docker进行本地开发 19
1.9.1　做好准备 19
1.9.2　实战操作 19
1.10　将请求路由到服务 20
1.10.1　理论阐释 21

1.10.2　实战操作 21

第2章　边缘服务

2.1　导语 23
2.2　使用边缘代理服务器控制对服务的访问 23
　　2.2.1　操作说明 24
　　2.2.2　实战操作 25
2.3　通过边车模式扩展服务 28
　　2.3.1　理论阐释 28
　　2.3.2　实战操作 29
2.4　使用API网关将请求路由到服务 31
　　2.4.1　设计时需要考虑的问题 32
　　2.4.2　实战操作 33
2.5　使用Hystrix停止级联故障 40
　　2.5.1　理论阐释 41
　　2.5.2　实战操作 41
2.6　速率限制 45
　　2.6.1　理论阐释 45
　　2.6.2　实战操作 46
2.7　使用服务网格解决共同关注的问题 46
　　2.7.1　理论阐释 47
　　2.7.2　实战操作 47

第3章　服务间通信

3.1　导语 49
3.2　从服务到服务的通信 50
　　3.2.1　理论阐释 50
　　3.2.2　实战操作 51
3.3　并发异步请求 56
　　3.3.1　理论阐释 56
　　3.3.2　实战操作 57
3.4　使用服务发现来查找服务 61
　　3.4.1　理论阐释 61
　　3.4.2　实战操作 62

3.5 服务器端负载均衡 67
　　3.5.1 理论阐释 67
　　3.5.2 实战操作 68
3.6 客户端负载均衡 69
　　3.6.1 理论阐释 69
　　3.6.2 实战操作 69
3.7 构建事件驱动的微服务 71
　　3.7.1 理论阐释 72
　　3.7.2 实战操作 72
3.8 不断演变的 API 77
　　3.8.1 理论阐释 77
　　3.8.2 实战操作 78

第 4 章　客户端模式 79
4.1 导语 79
4.2 使用依赖性的 Future 对并发进行建模 79
　　4.2.1 理论阐释 80
　　4.2.2 实战操作 80
4.3 服务于前端的后端 88
　　4.3.1 理论阐释 88
　　4.3.2 实战操作 90
4.4 使用 JSON 和 HTTP 实现 RPC 一致性 97
　　4.4.1 理论阐释 98
　　4.4.2 实战操作 98
4.5 使用 Thrift 103
　　4.5.1 理论阐释 103
　　4.5.2 实战操作 103
4.6 使用 gRPC 107
　　4.6.1 理论阐释 107
　　4.6.2 实战操作 107

第 5 章　可靠性模式 113
5.1 导语 113
5.2 使用断路器实现背压 114

5.2.1　理论阐释 114
　　　5.2.2　实战操作 115
　5.3　使用指数退避算法重试请求 126
　　　5.3.1　理论阐释 126
　　　5.3.2　实战操作 127
　5.4　通过缓存提高性能 130
　　　5.4.1　理论阐释 130
　　　5.4.2　实战操作 131
　5.5　通过 CDN 提供更高效的服务 136
　　　5.5.1　理论阐释 137
　　　5.5.2　实战操作 138
　　　5.5.3　优雅地降低用户体验 138
　5.6　通过游戏日演习验证容错能力 139
　　　5.6.1　理论阐释 139
　　　5.6.2　先决条件 140
　　　5.6.3　实战操作 140
　　　5.6.4　游戏日演习的模板 141
　5.7　引入自动化混沌工程 142
　　　5.7.1　理论阐释 142
　　　5.7.2　实战操作 143

第 6 章　安全性 145
　6.1　导语 145
　6.2　身份验证微服务 146
　　　6.2.1　理论阐释 146
　　　6.2.2　实战操作 148
　6.3　确保容器安全 162
　　　6.3.1　理论阐释 162
　　　6.3.2　实战操作 162
　6.4　安全配置 163
　　　6.4.1　理论阐释 163
　　　6.4.2　实战操作 164
　6.5　安全日志记录 176

	6.6	基础架构即代码 176
		6.6.1 理论阐释 176
		6.6.2 实战操作 177

第 7 章　监控和可观察性 181

   7.1 导语 181

   7.2 结构化 JSON 日志记录 182

        7.2.1 理论阐释 182

        7.2.2 实战操作 182

   7.3 使用 StatsD 和 Graphite 收集度量值 186

        7.3.1 理论阐释 186

        7.3.2 实战操作 186

   7.4 使用 Prometheus 收集度量值 190

        7.4.1 理论阐释 190

        7.4.2 实战操作 191

   7.5 通过跟踪使调试更容易 194

        7.5.1 理论阐释 195

        7.5.2 实战操作 195

   7.6 出现问题时发出警报 197

        7.6.1 理论阐释 198

        7.6.2 实战操作 198

第 8 章　扩展 203

   8.1 导语 203

   8.2 使用 Vegeta 对微服务进行负载测试 203

        8.2.1 理论阐释 203

        8.2.2 实战操作 204

   8.3 使用 Gatling 对微服务进行负载测试 209

        8.3.1 理论阐释 209

        8.3.2 实战操作 209

   8.4 构建自动扩展集群 212

        8.4.1 理论阐释 212

        8.4.2 实战操作 212

第 9 章 部署微服务 215
9.1 导语 215
9.2 配置服务以在容器中运行 216
9.2.1 理论阐释 217
9.2.2 实战操作 217
9.3 使用 Docker Compose 运行多容器应用程序 218
9.3.1 理论阐释 218
9.3.2 实战操作 218
9.4 在 Kubernetes 上部署服务 220
9.4.1 理论阐释 220
9.4.2 实战操作 221
9.5 使用金丝雀部署方式测试版本 223
9.5.1 理论阐释 223
9.5.2 实战操作 224

第 1 章　单体架构应用程序分解

本章包含以下操作秘笈。
- ❑　组织开发团队以采用微服务架构。
- ❑　按业务功能分解微服务。
- ❑　识别有界上下文。
- ❑　迁移生产环境中的数据。
- ❑　重构单体架构应用程序。
- ❑　将单体架构应用程序升级为服务。
- ❑　升级测试套件。
- ❑　使用 Docker 进行本地开发。
- ❑　将请求路由到服务。

1.1　导　　语

俗话说"万事开头难",微服务(Microservices)最困难的就是起步阶段。许多团队都会发现自己不知不觉地就将应用程序构建为一个功能不断增长的、难以管理的单体架构(Monolithic)应用程序代码库,并且不知道如何将其分解为更容易管理的、可单独部署的服务。本章中的秘笈将说明如何实现从单体架构应用程序到微服务的过渡。许多秘笈都不会涉及任何代码。相反,它们将专注于架构设计以及如何以最佳方式组建团队以从事微服务开发。

我们将学习如何开始从单一的单体架构应用程序代码库过渡到微服务套件。当你开始使用这种新的架构风格开发应用程序时,还需要了解如何应对起步阶段的困难挑战。

1.2　组织开发团队

康威定律(Conway's Law)告诉我们,组织设计的系统架构是组织本身的沟通结构的副本,也就是说,产品的系统架构将打上开发它的组织的沟通结构的烙印。这通常意味着,软件工程团队的组织结构将对其开发的软件的设计结构产生深远的影响。

例如，当新的创业公司开始构建软件时，团队规模很小，有时可能只有一两名软件开发工程师。在这种情况下，开发工程师将需要处理所有的东西，包括前端和后端系统以及操作。单体架构就非常适合这种组织结构，它使得工程师可以在任何给定时间处理系统的任何部分的内容，而无须在代码库之间切换。

随着团队的成长和产品功能的增加，作为公司总裁，你开始意识到微服务的好处，这时可以考虑采用一种通常称为反向康威操纵（Inverse Conway Maneuver）的技术。该技术建议你改进团队和组织结构，以鼓励产生你想看到的那种产品架构。对于微服务来说，这通常指的是将开发工程师组织成小团队，这些小团队各司其职，仅负责少量相关服务。

作为公司总裁，你可以提前建立团队以适应这种结构，通过限制团队内部的沟通和决策开销来激励工程师构建服务。打个比方，单体架构模式就好像是一个人情社会，有什么事私下打招呼，托个人情就办了；而微服务模式就好像是一个法治社会，任何事情都要走正规流程，通过微服务处理。所以，当程序功能越来越复杂时，单体架构模式容易陷入混乱，不容易管理；而微服务模式按流程办事，功能扩展容易，维护管理也更加简单。

当然，单体架构模式也有其优势，简而言之，当添加某项功能作为服务时，如果其成本大于向单体架构程序中添加功能的成本，那么单体架构就有继续存在的价值。在这种情况下，以单体架构方式组织团队可以减少开发服务的成本。

本秘笈针对有能力变更组织结构的公司总裁和其他领导者。

1.2.1 实战操作

重组团队绝非易事，要考虑许多不容忽视但又不易发现的因素。诸如开发人员个性、个人强项和弱点，以及过去的历史等因素固然不在本秘笈的讨论范围之内，但是在进行任何更改时都应该仔细考虑。

此秘笈中的步骤提供了一种从围绕单体架构代码库的组织转变为针对微服务进行优化的团队可能方法，但是请记住，每家公司或组织都有具体情况，并没有一种"放之四海而皆准"的万能解决方案。

以下步骤仅供参考。

（1）与组织中的其他利益相关者合作，制定产品路线图。对于公司或组织在短期内将面临的挑战，你获得的信息可能会有遗漏，但是只要尽力搜集即可。通常而言，路线图上的短期项目应该非常详细，而长期项目则是一般指导性的。

（2）使用产品路线图，尝试确定你需要的技术功能（这些功能影响到你给用户提供的价值）。例如，你可能打算使用严重依赖搜索的功能，或者可能还具有许多依赖内容上传和管理的功能。这意味着搜索和上传是你需要投资的两项技术功能。

（3）当你在路线图上看到自己想要的模式出现时，请尝试确定你的应用程序的主要功能区域，并注意预估需要在每个区域中投入的工作量。对于那些你期望在短期或中期见效的功能领域，则需要分配更高的优先级和投注更多资源。

（4）创建新团队，最好由 4~6 名工程师组成，他们负责应用程序中的功能区域之一。从你预计的在下一个季度需要投入最多资源的功能区域开始。这些团队可以专注于后端服务，也可以是包括移动和 Web 工程师在内的跨职能团队。拥有跨职能团队的好处在于，团队可以自动交付应用程序的整个垂直组件。服务工程师与使用其服务的工程师的结合也将实现更多的信息共享，并可能产生良好的无缝协作。

1.2.2 示例讨论

使用这种方法时，最终应该由小型、团结且专注的团队来负责应用程序的核心领域。团队的本质是，团队中的个体应该开始明白创建单独管理和部署的代码库的好处，他们可以自主地工作，而不会产生与其他团队协调变更和部署的高昂开销。

为了更好地理解这些步骤，不妨想象一下，你的公司开发了一款图像消息应用程序。该应用程序允许用户使用智能手机拍照，并将照片与消息一起发送给联系人列表中的朋友。他们的朋友也可以给他们发送带有消息的照片。该虚构产品的虚构路线图可能涉及需要增加对短视频、照片滤镜和表情符号的支持。你现在就应该知道，录制、上传和播放视频的能力，应用照片滤镜的能力以及发送富文本格式的能力对你的公司和产品至关重要。此外，根据经验你也应该知道，用户需要注册、登录和维护好友列表。

根据上述例，你可以决定将开发工程师组织成以下 3 个小团队。

- 一个媒体团队，负责上传、处理和播放视频，滤镜以及存储和交付。
- 一个消息传递团队，负责发送带有相关文本的照片或视频消息。
- 一个用户团队，负责提供可靠的身份验证、注册、登录和社交功能。

1.3 按业务功能分解微服务

在产品开发的早期阶段，单体架构最适合于尽可能快速、简单地向用户交付功能。这是正常的，因为在产品开发这个阶段，你不存在因为客户流量暴增而必须扩展团队、代码库和服务能力这样的"奢侈"问题。按照良好的设计规范，你可以将应用程序的重点放在易于阅读的模块化代码模式上，这样做可以使工程师自主地处理代码的不同部分，并尽可能地防止当将分支代码合并到主代码中并进行部署时产生的合并冲突。

1.3.1 理论阐释

微服务架构要求你比单体架构中已经遵循的良好设计规范更进一步。要围绕微服务组织小型自治团队，你应该考虑首先确定应用程序提供的核心业务功能。业务功能（Business Capability）是商学院的一个术语，它描述了公司或组织创造价值的各种方式。例如，你的内部订单管理系统将负责处理客户订单；如果你有一个社交应用程序，允许用户提交由用户生成的内容（如照片），则你需要一个照片上传系统来提供这种业务功能。

在考虑系统设计时，业务功能与面向对象设计（Object-Oriented Design，OOD）中的单一职责原则（Single Responsibility Principle，SRP）密切相关。单一职责原则是面向对象设计的 5 个基本原则之一，它规定一个类应该只有一个发生变化的原因。微服务本质上就是将单一职责原则（SRP）扩展到了代码库。如果你能理解这一点，那么它将有助于你设计大小适当的微服务。

服务应该有一项主要的任务，并且应该将它完成好。在上述示例中，服务可能是存储图像和应用滤镜，也可能是传递消息，或是创建和验证用户账户。

1.3.2 实战操作

按业务功能分解单体架构应用程序是一个过程。对于确定需要的每一项新服务，都可以并行执行这些步骤，当然你也可以考虑从某一项服务开始，然后将学习到的分解经验应用于后续的服务中。

（1）确定你的单体架构应用程序当前提供的业务功能。这将是我们第一项服务的目标。理想情况下，这种业务功能应该是你在上一个秘笈所研究的路线图中设定的，并且该业务功能的所有权可以授予你新创建的团队之一。

仍以前述虚构的图像消息应用程序为例，假设在路线图上我们将上传和显示媒体的业务功能作为已确定的第一项服务。该功能目前是作为单个模块和控制器实现的，在 Ruby on Rails 单体架构中的表示如图 1-1 所示。

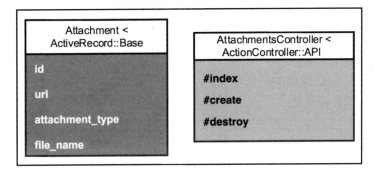

图 1-1

（2）从图 1-1 中可以看到，AttachmentsController 有 4 个方法——在 Ruby on Rails 术语中，这些方法被称为操作（Action），这 4 个方法大致与你要在 Attachment（附件）资源上执行的创建/检索/更新/删除（Create/Retrieve/Update/Delete，CRUD）操作相对应。更新操作严格来说是不需要的，因此可以忽略它。这些操作可以很好地映射到 RESTful 服务中，因此你可以使用以下 API 设计、实现和部署微服务：

```
POST /attachments
GET /attachments/:id
DELETE /attachments/:id
```

（3）在部署了新的微服务后（后续的秘笈将讨论迁移数据），现在可以开始修改客户端代码路径以使用新服务。你可以从替换 AttachmentsController 操作方法中的代码开始，向新的微服务发出 HTTP 请求。在 1.7 节 "将单体架构应用程序升级为服务" 秘笈中将介绍执行此操作的技术。

1.4 识别有界上下文

在设计微服务时，一个常见的容易让人混淆的问题是服务的大小。这种混乱会导致开发工程师专注于诸如在某个特定服务中的代码行数之类的事情。事实上，代码行的多寡是衡量软件的糟糕指标。更有用的指标是从服务提供的业务功能和它帮助管理的领域对象两个方面来考虑聚焦服务所扮演的角色。

1.4.1 理论阐释

在设计微服务时，应该设计与其他服务的耦合度较低的服务，这样的话，当在产品中引入新功能或对现有功能进行更改时，就不必修改太多的内容。总而言之，我们希望服务仅承担单一的责任。

分解单体架构应用程序时，常见的实用方式是查看数据模块，这对于决定要提取哪些服务很有帮助。仍以虚构的图像消息应用程序为例，我们可以想象该应用程序使用了如图 1-2 所示的数据模块。

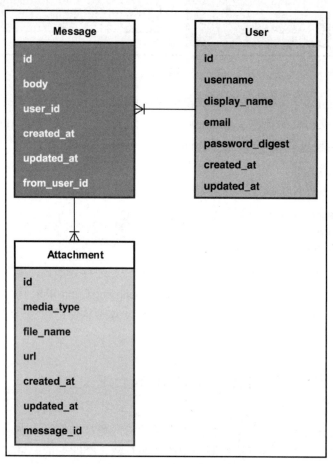

图 1-2

可以看到，图 1-2 有以下 3 张表。

- Message 表：用于消息。
- User 表：用于用户。
- Attachment 表：用于附件（指发送的图片和视频等）。

User 实体与 Message 实体具有一对多关系，即每个用户都可以发出多条消息，或接收到多条消息。

Message 实体与 Attachment 实体也有一对多关系，即每条消息都可以包含多个附件。

随着应用程序的发展以及用户要求的功能越来越丰富，我们会发现上述数据模块越来越不敷使用。例如，上述数据模块则不包含有关社交关系的任何内容。假设我们希望某个用户能够关注其他用户，那么这应该被定义为一种非对称关系，即用户 1 可以关注用户 2，但这并不意味着用户 2 一定要关注用户 1。

有很多方法可以为这种关系建模。我们将重点介绍最简单的一项，即邻接表（Adjacency List）。下面来看图 1-3。

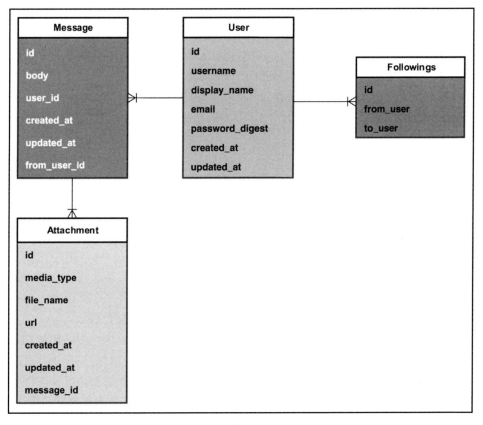

图 1-3

现在我们多了一个实体 Followings，它表示两个用户之间的关注关系。这在单体架构应用程序中是非常有效的，但是在微服务模式下则会产生问题。如果我们要构建两个新服务，一个用于处理附件，另一个用于处理社交关系图（因为它们是两个不同的职责），那么我们现在就有了两个用户定义。这种模块重复通常是必要的。如果不想要这种重复，则替代方法是让多个服务访问并更新同一模块，而这是非常脆弱的，并且很可能会导致代码不可靠。

在这种情况下，有界上下文（Bounded Context）可以帮忙解决问题。有界上下文是域驱动设计（Domain-Driven Design，DDD）中的一个术语，DDD 通过将大型模块划分为不同的有界上下文并明确其相互关系来处理大型模块，它定义了特定模块有意义的系统区域。在上述示例中，社交关系图服务会有一个 User 模块，其有界上下文将是用户的社交关系图（这很容易处理）。媒体服务也会有一个 User 模块，其有界上下文将是用户的照片和视频。

识别这些有界上下文非常重要，尤其是在解构单体架构应用程序时。你可能会发现，随着单体架构应用程序代码库的增长，先前讨论的业务功能（上传和查看照片、视频、用户关系等）可能会共享相同的不断膨胀的 User 模块，最后不得不将其拆解开。这可能是一个需要技巧但是颇有启发性的重要过程。

1.4.2 实战操作

决定如何定义系统内的有界上下文是一项有意义的工作。该过程本身将促使团队就系统中的模块，以及各种系统之间必须发生的各种交互进行许多有趣的讨论。

下面来看以下操作。

（1）在团队可以开始定义其使用的有界上下文之前，应该先列出其工作的系统各部分所拥有的模块。

例如，媒体团队显然将拥有 Attachment 模块，但他们也需要拥有与用户以及消息相关的信息。Attachment 模块可以完全在媒体团队自己的服务上下文中被维护，但其他服务则必须具有定义明确的有界上下文，必要时可以与其他团队进行交流。

（2）一旦团队确定了潜在的共享模块，就最好与使用相似模块或相同模块的其他团队进行讨论。

（3）在这些讨论中，应敲定模块的界限，并确定是共享模块的实现（在微服务架构中，这需要进行服务到服务的调用），还是采用不同的方式来开发和维护独立的模块实现。如果选择开发单独的模块实现，那么清楚定义该模块适用的有界上下文将变得很重要。

（4）团队应从自身的角度明确说明模块的边界、应用程序的特定部分或应使用模块的特定代码库。

1.5 迁移生产环境中的数据

单体架构应用程序的代码库通常使用主关系数据库来实现持久保存。当前 Web 框架通常与对象关系映射（Object-Relational Mapping，ORM）打包在一起，它允许你使用与数据库中的表相对应的类来定义域对象。

1.5.1 理论阐释

模块类的实例对应于数据库表中的行。随着单体架构应用程序代码库的增长，添加其他数据存储（如文档或键值存储）的情况也并不稀奇。

微服务不应与单体架构应用程序共享访问权限，并连接到相同的数据库。这样做在尝试协调数据迁移（如模式更改）时不可避免地会引起问题。当你在一个代码库中更改数据的写入方式，而在另一代码库中数据的读取方式并未发生变化时，即使是无模式的存储也将导致问题。由于这个原因以及其他一些原因，最好让微服务完全管理它们自己的数据存储。

从单体架构应用程序过渡到微服务时，制定迁移数据的策略非常重要。很多时候，一个团队会提取微服务的代码并留下数据，以备不时之需。除了难以管理迁移，单体架构应用程序关系数据库中的故障现在还将对服务产生级联影响，从而导致难以调试的生产环境下的事件。

有一种用于管理大规模数据迁移的流行技术是设置双重写入。部署新服务后，你将拥有两条写入路径：一条写入路径是从原来的单体架构应用程序代码库到其数据库；另一条写入路径则是从新服务到它自己的数据存储。

你应该确保写入能够到达这两条代码路径。现在，你将从新服务投入生产环境中的那一刻起开始复制数据，从而使你可以使用脚本或类似的脱机任务回填旧数据。将数据写入两个数据存储中后，就可以修改所有不同读取路径。无论代码用于直接查询单体架构应用程序数据库的何处，都应使用对新服务的调用来替换查询。修改完所有读取路径后，即可删除仍在写入旧位置中的所有写入路径。最后，你可以安全删除旧数据（当然，别忘记保留备份）。

1.5.2 实战操作

在向微服务过渡时，将数据从单体架构应用程序的数据库迁移到新服务的新存储中，而不影响数据的可用性或一致性,这是一项困难但又很常见的任务。

以我们虚构的图像消息应用程序为例,可以想象一个场景,我们想要创建一个新的微服务来处理媒体上传。在这种情况下,我们将遵循一种常见的双重写入模式,操作如下。

(1)在编写用于处理媒体上传的新服务之前,我们假设单体架构应用程序的结构看起来如图 1-4 所示。其中,单体架构应用程序正在处理 HTTP 请求,它基本上是将 Multipart/Form 编码的内容作为二进制对象读取,并将文件存储在分布式文件存储(例如,Amazon 的 S3 服务)中,然后将有关文件的元数据(Metadata)写入称为 attachments 的数据库表中。

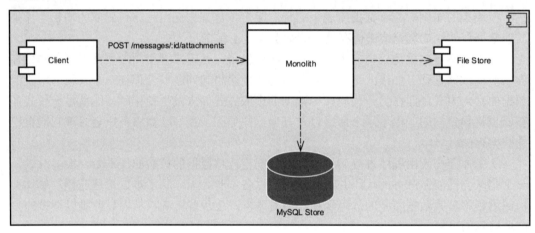

图 1-4

原　文	译　文
Client	客户端
Monolith	单体架构应用程序
MySQL Store	MySQL 数据库存储
File Store	文件存储

(2)在编写新服务后,你现在有两个写入路径。在单体架构应用程序的写入路径中,将调用你的服务,以便将数据同时写入单体架构应用数据库和新服务所使用的数据库中。现在,你可以复制新数据,并且可以编写脚本来回填旧数据。你的架构现在看起来应如图 1-5 所示。

(3)在客户端和单体架构应用程序代码中,找到所有的读取路径,并更新它们以使用新服务。现在,所有读取将转到你的服务中,这将能够给出一致的结果。

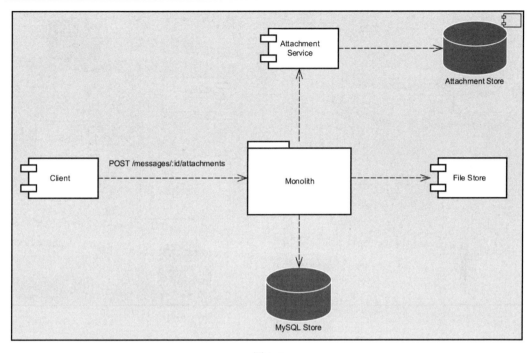

图 1-5

原　　文	译　　文
Client	客户端
Monolith	单体架构应用程序
MySQL Store	MySQL 数据库存储
File Store	文件存储
Attachment Service	Attachment 服务
Attachment Store	Attachment 存储

（4）在客户端单体架构应用程序代码中找到所有的写入路径，并更新它们以使用新服务。现在，所有读写操作都将转移到你的服务中，你可以安全地删除旧数据和代码路径。你的最终架构应类似于图 1-6。

仔细阅读的读者可能会发现图 1-6 新增了边缘代理（Edge Proxy），别着急，在后续章节中将会对此进行详细讨论。

使用这种方法和步骤，你就可以安全地将数据从单体架构应用程序数据库迁移到为新的微服务创建的新存储中，而无须将生产环境停机。重要的是，不要跳过步骤（4）；否则你将无法真正意识到微服务架构的好处，反而可能感受到双重写入的麻烦。

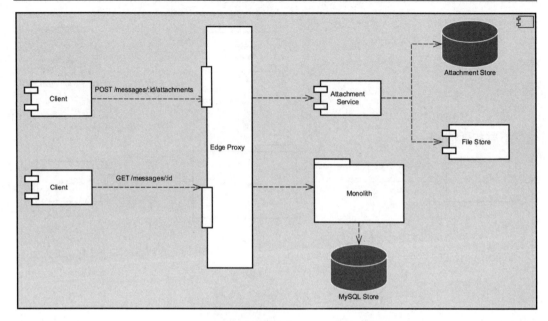

图 1-6

原　　文	译　　文
Client	客户端
Monolith	单体架构应用程序
MySQL Store	MySQL 数据库存储
File Store	文件存储
Attachment Service	Attachment 服务
Attachment Store	Attachment 存储
Edge Proxy	边缘代理

1.6　重构单体架构应用程序

当过渡到微服务架构中时，常见的错误是忽略原有的单体架构应用程序，而只是将新功能作为一项服务来构建。

1.6.1　理论阐释

当团队认为单体架构应用程序已经失控，并且代码变得一团乱麻，为新功能进行修改还不如重起炉灶时，就会发生想要忽略原有的单体架构应用程序的情况。这可能特别有吸引力，因为构建没有遗留包袱的绿地代码的想法听起来比重构脆弱的陈旧代码好得多。

💡 **提示：**

棕地（Brownfield）这个概念就是源于它字面上的意思，即棕色的土地。棕地是指在城市不断扩张过程中留下的大量工业旧址，它们往往被闲置荒芜，并且可能含有危害性物质和污染物，导致再利用变得非常困难。与棕地相对应的概念是绿地（Greenfield），绿地一般指未被用于开发和建设并覆盖有绿色植物的土地。

软件开发领域借用了棕地和绿地的概念，绿地技术指的是在全新环境中从头开发的软件项目；而棕地技术指的是在遗留系统之上开发和部署新的软件系统，或者需要与已经在使用的其他软件共存。

但是，作为公司总裁，你应该对完全放弃单体架构应用程序的诱惑说"不"。为了成功地按业务功能分解单体架构应用程序，并将其发展为一组完善的、单一职责的微服务，你需要确保单体架构应用程序的代码库的格式良好，然后进行仔细的分解和测试；否则，你将获得大量新服务，但这些服务却无法对你的域进行清晰的建模（因为它们与单体架构应用程序中的功能重叠），并且你将无法继续使用单体架构应用程序中已有的任何代码。这样你的用户会因为体验很差而感到不开心，甚至大量流失。更糟糕的是，随着技术上的负担变得越来越难以承受，你的团队也将很可能会疲于奔命，斗志全无。

因此，我们必须采用完善而可靠的设计原则，采取持续而积极的步骤来重构单体架构应用程序。关于重构的问题，当前已经有许多很棒的书，例如 Martin Fowler 编写的 *Refactoring*（重构）和 Michael Feathers 编写的 *Working Effectively with Legacy Code*（有效处理遗留代码），但最重要的是要知道，重构绝不是一项可以一劳永逸的工作。很少有产品团队或公司可以奢侈地将所有业务都停掉，然后耐心等待工程团队进行重构，并花费时间使他们的代码更易于更改。事实上，尝试这样做的工程团队也很少会成功。重构只能是一个持续而可靠的过程。即一边修改一边保持业务的正常运转。

无论你的团队如何安排其工作，都有必要保留适当的时间进行重构。一个指导原则是，每当你要进行更改时，首先要使得更改容易进行，然后实施更改。你的目标是使你的单体架构应用程序的代码更易于使用、更易于理解并且不那么脆弱。此外，你还应该能够开发出将派上用场的强大测试套件。

一旦单体架构应用程序结构完好，就可以不断抽离分解出服务，使得单体架构应用程序的结构持续收缩。大多数单体架构应用程序代码库的另一方面是为通过浏览器提供的动态生成的视图和静态资产提供服务。如果你的单体架构应用程序对此负责，则应考虑将 Web 应用程序组件移至单独提供的 JavaScript 应用程序中。这将允许你从多个方向上收缩单体架构应用程序。

1.6.2 实战操作

重构任何代码库都是一个持续的过程。对于单体架构应用程序而言，有些技术可以很好地工作。在本章示例中，我们将详细说明可简化重构 Ruby on Rails 代码库的步骤，具体如下。

（1）使用前述秘笈中描述的技术，确定应用程序中的业务功能和有界上下文。让我们专注于上传图片和视频的能力。

（2）在 controllers、models 和 views 旁边创建一个名为 app/services 的目录。该目录将包含你的所有服务对象。服务对象是许多 Rails 应用程序中使用的一种模式，用于将概念服务分解为不继承任何 Ruby on Rails 功能的 ruby 对象。这使得将封装在服务对象中的功能迁移到单独的微服务中变得更加容易。没有一种方法可以构造服务对象。笔者更喜欢让每个对象代表一个服务，并将希望该服务负责的操作作为方法移动到该服务对象中。

（3）在 app/services 下创建一个名为 attachments_service.rb 的新文件，并为其指定以下定义：

```ruby
class AttachmentsService

  def upload
    # ...
  end

  def delete!
    # ...
  end

end
```

（4）查看 app/controllers/attachments_controller.rb 文件中 AttachmentsController#create 方法的源代码，它当前负责创建 Attachment 实例并将文件数据上传到附件存储（在本示例中，这个附件存储是 Amazon S3 存储桶）中。以下是我们移至新创建的服务对象中所需的功能：

```ruby
# POST /messages/:message_id/attachments
def create
  message = Message.find_by!(params[:message_id], user_id:
```

```
    current_user.id)
  file = StorageBucket.files.create(
      key: params[:file][:name],
      body: StringIO.new(Base64.decode64(params[:file][:data]), 'rb'),
      public: true
  )
  attachment = Attachment.new(attachment_params.merge!(message: message))
  attachment.url = file.public_url
  attachment.file_name = params[:file][:name]
  attachment.save
  json_response({ url: attachment.url }, :created)
end
```

（5）打开 app/services/attachments_service.rb 文件中新创建的服务对象，并将文件上传的责任移至 AttachmentsService#upload 方法中，具体如下：

```
class AttachmentsService
  def upload(message_id, user_id, file_name, data, media_type)
    message = Message.find_by!(message_id, user_id: user_id)
    file = StorageBucket.files.create(
        key:    file_name,
        body:   StringIO.new(Base64.decode64(data), 'rb'),
        public: true
    )
    Attachment.create(
        media_type: media_type,
        file_name:  file_name,
        url:        file.public_url,
        message:    message
    )
  end

  def delete!
  end
end
```

（6）现在，上传 app/controllers/attachments_controller.rb 中的 AttachmentsController#create 方法，以使用新创建的 AttachmentsService#upload 方法，具体如下：

```
# POST /messages/:message_id/attachments
def create
  service = AttachmentService.new
```

```
    attachment = service.upload(params[:message_id], current_user.id,
        params[:file][:name], params[:file][:data],
        params[:media_type])
    json_response({ url: attachment.url }, :created)
end
```

（7）对 AttachmentsController#destroy 方法中的代码重复此过程，将其职责移至新的服务对象中。完成之后，AttachmentsController 中的任何代码都不应直接与 Attachments 模块进行交互。相反，它应该通过 AttachmentsService 服务对象执行其功能。

现在，你已经隔离了管理单个服务类附件的责任。该类应封装所有业务逻辑，这些业务逻辑最终将被移至新的附件服务中。

1.7 将单体架构应用程序升级为服务

从单体架构应用程序过渡到服务最复杂的方面之一就是请求的路径设置。在后续秘笈和章节中，我们将探讨如何将你的服务公开到 Internet 中，这样移动和 Web 客户端应用程序就可以直接与它们通信。但是，就目前阶段而言，你仍然需要让单体架构应用程序充当路由，这是一个很实用的中间步骤安排。

1.7.1 理论阐释

当你将单体架构应用程序拆分为可维护的很小的微服务时，可以将单体架构应用程序中的代码路径替换为对服务的调用。根据用于构建单体架构应用程序的编程语言或框架，这些代码段可以被称为控制器操作、视图或其他内容。我们将继续假设你的单体架构应用程序是在流行的 Ruby on Rails 框架中构建的；在这种情况下，我们将研究的就是控制器的操作。我们还将假定你已经开始重构单体架构应用程序，并已按照 1.6 节的秘笈中的描述创建了一个或多个服务对象。

在遵循最佳做法时，这一点很重要。在后续章节中，我们将介绍诸如断路器之类的概念，这些概念在进行从服务到服务（Service-to-Service）的通信时变得非常重要。现在，请注意，从你的单体架构应用程序到服务的 HTTP 调用可能会失败，并且你应该考虑如何以最佳方式处理这种情况。

1.7.2 实战操作

本秘笈需要执行以下操作。

（1）打开我们在上一个秘笈中创建的服务对象。我们将修改服务对象，以便能够调用负责管理附件的外部微服务。为简单起见，我们将使用 Ruby 标准库中提供的 HTTP 客户端。该服务对象应位于 app/services/attachments_service.rb 文件中，具体如下：

```
class AttachmentsService

    BASE_URI = "http://attachment-service.yourorg.example.com/"

    def upload(message_id, user_id, file_name, data, media_type)
        body = {
            user_id: user_id,
            file_name: file_name,
            data: StringIO.new(Base64.decode64(params[:file]
            [:data]), 'rb'),
            message: message_id,
            media_type: media_type
        }.to_json
        uri = URI("#{BASE_URI}attachment")
        headers = { "Content-Type" => "application/json" }
        Net::HTTP.post(uri, body, headers)
    end

end
```

（2）打开位于 pichat/app/controllers/ 中的 attachments_controller.rb 文件，然后查看以下创建操作。由于在 1.6 节"重构单体架构应用程序"中已经完成了重构工作，因此现在只需要进行少量更改即可使控制器与我们的新服务对象一起工作，具体如下：

```
class AttachmentsController < ApplicationController
    # POST /messages/:message_id/attachments
    def create
        service = AttachmentService.new
        response = service.upload(params[:message_id], current_user.id,
            params[:file][:name],
            params[:file][:data],
            params[:media_type])
        json_response(response.body, response.code)
    end
    # ...
end
```

1.8 升级测试套件

拥有一个良好的测试套件将极大地帮助你从单体架构应用程序转变为微服务。每次你从单体架构应用程序代码库中删除功能时，都需要进行更新测试以确保功能正常。你可以尝试用对服务进行外部网络调用的测试来取代 Rails 应用程序中的单元测试和功能测试，但是这种方法有很多缺点。进行外部调用的测试很容易因间歇性的网络连接问题而导致失败，并且这样的测试还需要花费大量时间。

因此，我们可以不进行外部网络的调用，而转为修改单体架构应用程序测试，通过使用桩代码（Stub）来伪造一个方法伪装微服务，进而阻断对实际微服务的调用。在使用桩代码之后，对微服务调用的测试就不再脆弱，并且运行速度更快。只要你的微服务满足你开发的 API 接口，测试就将成为单体架构应用代码库运行状况的可靠指标。对微服务进行向后不兼容的更改是另一个主题，将在以后的秘笈中介绍。

1.8.1 做好准备

在测试中，我们将使用 webmock gem 对外部 HTTP 请求进行桩代码处理，因此你需要更新单体架构应用程序的 gemfile，以将 webmock gem 包含在测试组中，具体如下：

```
group :test do
    # ...
    gem 'webmock'
end
```

你还应该更新 spec/spec_helper.rb 以禁用外部网络请求。这将使你在编写其余测试代码时保持诚实，具体如下：

```
require 'webmock/rspec'
WebMock.disable_net_connect!(allow_localhost: false)
```

1.8.2 实战操作

现在，你的项目中已包含 webmock，你可以开始在规范中使用桩代码处理 HTTP 请求。再次打开 specs/spec_helper.rb 并添加以下内容：

```
stub_request(:post, "attachment-service.yourorg.example.com").
    with(body: {media_type:1}, headers: {"Content-Type" => /image\/.+/}).
    to_return(body: { foo: bar })
```

1.9 使用 Docker 进行本地开发

如前文所述，微服务的作用是解决一组特定的问题，但是这也引入了一些新的挑战。团队中的工程师可能会遇到的一项挑战是进行本地开发。在使用单体架构应用程序的情况下，基本上没有移动部件需要进行管理——通常而言，你只需在工作站上运行数据库和应用程序服务器即可完成工作。但是，随着你开始创建新的微服务，情况变得更加复杂。

使用容器是管理这种复杂性的好方法。Docker 是一个流行的开源软件容器化平台。Docker 允许你指定如何将应用程序作为容器运行——这是一种轻量级的标准化部署单元。有关 Docker 的书籍和在线说明文档都很多，所以在这里我们不再赘述，你只要知道容器封装了运行应用程序所需的所有信息即可。如前文所述，单体架构应用程序通常至少需要一台应用程序服务器和一台数据库服务器，它们都将在各自的容器中运行。

Docker Compose 是用于运行多容器应用程序的工具。Compose 允许你在 YAML 配置文件中定义应用程序容器。使用此文件中的信息，即可构建和运行你的应用程序。Compose 将在单独的容器中管理配置文件中定义的所有各种服务，使你可以在工作站上运行复杂的系统以进行本地开发。

1.9.1 做好准备

在按照本秘笈中的步骤进行操作之前，你需要安装一些必备软件。

（1）安装 Docker。从 Docker 网站下载安装软件包并按照说明进行操作。其网址如下：

https://www.docker.com/docker-mac

（2）通过在 macOS X 上执行以下命令行来安装 docker-compose：

```
brew install docker-compose
```

在 Ubuntu Linux 上，可以执行以下命令行：

```
apt-get install docker-compose
```

安装了上述两个软件包后，可以按照本秘笈中的步骤进行操作。

1.9.2 实战操作

本秘笈需要执行以下操作。

（1）在 Rails 应用程序的根目录中，创建一个名为 Dockerfile 的文件，其内容如下：

```
FROM ruby:2.3.3
RUN apt-get update -qq && apt-get install -y build-essential
libpq-dev nodejs
RUN mkdir /pichat
WORKDIR /pichat
ADD Gemfile /pichat/Gemfile
ADD Gemfile.lock /pichat/Gemfile.lock
RUN bundle install
ADD . /pichat
```

（2）创建一个名为 docker-compose.yml 的文件，其内容如下：

```
version: '3'
services:
    db:
        image: mysql:5.6.34
        ports:
            - "3306:3306"
        environment:
            MYSQL_ROOT_PASSWORD: root

    app:
        build: .
        environment:
            RAILS_ENV: development
        command: bundle exec rails s -p 3000 -b '0.0.0.0'
        volumes:
            - .:/pichat
        ports:
            - "3000:3000"
        depends_on:
            - db
```

（3）通过运行 docker-compose up app 命令启动应用程序。你应该能够通过在浏览器中输入 http://localhost:3000/ 来访问你的单体架构应用程序。可以将这种方法用于所编写的新服务中。

1.10 将请求路由到服务

在前述秘笈中，我们专注于使你的单体架构应用程序将请求路由到服务。这项技术是一个很好的开始,因为它不需要更改客户端即可工作。客户端仍然向单体架构应用程序发出请求,而单体架构应用程序则通过其控制器操作将请求封送(Marshal)给微服务。但是,当项目进行到某个阶段时,要真正受益于微服务架构,仍然必须从关键路径中移除单体架构应用程序(也就是所谓的"没有中间商赚差价"),并允许客户端直接向微服务发出请求。

1.10.1 理论阐释

在转换微服务架构时,比较常见的做法是,工程师直接将其组织的第一个微服务直接公开到 Internet(通常会使用不同的主机名)中。但是,随着你开发更多的服务,并且在监视、安全性和可靠性方面需要一定程度的一致性时,这将变得难以管理。

一般来说,面向 Internet 的系统面临许多挑战,它们需要能够处理许多与安全相关的问题、速率限制,以及流量的周期性峰值等。要为你公开到公共 Internet 中的每项服务都执行此操作,那么成本将是非常高的,并且还需要非常快速。因此,你应该考虑开发一个单一的边缘服务,让该边缘服务支持从公共 Internet 到内部服务的路由请求。良好的边缘服务应支持通用功能,如动态路径重写、卸除负载和身份验证等。幸运的是,当前有许多优秀的开源边缘服务解决方案。在本秘笈中,我们将使用一个名为 Zuul 的 Netflix 项目。

1.10.2 实战操作

本秘笈需要执行以下操作。

(1)我们需要创建一个名为 Edge Proxy 的新 Spring Boot 服务,其主类为 EdgeProxyApplication。

(2)Spring Cloud 包含一个嵌入式 Zuul 代理。可以通过将@EnableZuulProxy 注解添加到 EdgeProxyApplication 类中来启用它:

```java
package com.packtpub.microservices;

import org.springframework.boot.SpringApplication;
import org.springframework.boot.autoconfigure.SpringBootApplication;
import org.springframework.cloud.netflix.zuul.EnableZuulProxy;

@EnableZuulProxy
@SpringBootApplication
public class EdgeProxyApplication {
```

```
    public static void main(String[] args) {
        SpringApplication.run(EdgeProxyApplication.class, args);
    }
}
```

（3）在 src/main/resources/ 下创建一个名为 application.properties 的文件，其内容如下：

```
zuul.routes.media.url = http://localhost:8090
ribbon.eureka.enabled = false
server.port = 8080
```

在上述代码中，它告诉 zuul 将对 /media 的请求路由到运行在端口 8090 上的服务。我们将在后续章节讨论到服务发现时，再详细介绍这个 eureka 选项，现在只需确保将其设置为 false 即可。

此时，你应该能够通过 zuul 代理将请求路由到适当的服务。这是你向成功构建微服务架构迈出的最大步骤之一。恭喜你！

第 2 章 边缘服务

本章包含以下操作秘笈。
- ❏ 使用边缘代理服务器控制对服务的访问。
- ❏ 通过边车模式扩展服务。
- ❏ 使用 API 网关将请求路由到服务。
- ❏ 使用 Hystrix 停止级联故障。
- ❏ 速率限制。
- ❏ 使用服务网格解决共同关注的问题。

2.1 导　语

经过本书第 1 章"单体架构应用程序分解"的学习之后，相信你已经拥有了将单体架构应用程序分解为微服务的经验。你应该了解，可能会有许多问题出现在单体架构应用程序或服务代码库本身之外。

例如，将服务公开到 Internet 中、控制路由和建立弹性等，这些都是可以通过所谓边缘服务（Edge Services）解决的问题。这些服务存在于架构的边缘，通常用于处理来自公共 Internet 中的请求。幸运的是，由于很多问题都是公共的，因此在这方面出现了一些开源项目，可以帮助我们处理大多数挑战。本章将使用许多出色的开源软件。

通过本章中的秘笈，你将学习到如何使用开源软件将服务公开到公共 Internet 中、控制路由、扩展服务的功能以及处理在部署和扩展微服务时会遇到的许多常见挑战。你还将学习如何针对服务进行客户端开发，使服务变得更轻松，以及如何实现微服务架构监控和可观察性的标准化。

2.2 使用边缘代理服务器控制对服务的访问

在第 1 章"单体架构应用程序分解"中，我们修改了单体架构应用程序的代码库，以轻松路由到我们的微服务。这种方法是可行的，因为只需要做很少的修改，就可以使单体架构应用程序成为一个中介。但是，这样简化处理的结果，就是最终你的单体架构应用程序将成为架构开发和弹性的瓶颈。当你尝试扩展服务并构建更多微服务时，每次对服务进行 API 更改都需要更新和重新部署单体架构应用程序。此外，你的单体架构应用程序将不得不处理与服务的连接，并且可能没有正确配置以处理边缘问题（如负载卸除或断路）。在 1.10 节"将请求路由到服务"中，我们介绍了边缘代理（Edge Proxy）的概念。使用边缘代理服务器将服务公开到公共 Internet 中，可帮助你处理大部分公开服务必须解决的共同问题。诸如请求路由、负载卸除、背压（Back Pressure）和身份验证之类的要求都可以在单个边缘代理层中处理，而不必由公开到 Internet 中的每项服务重复执行。

边缘代理是位于基础结构边缘的代理服务器，可提供对内部服务的访问。你可以将边缘代理视为内部服务架构的"前门"，它允许 Internet 中的客户端向你部署的内部服务发出请求。当前已经有多个包含强大功能集的开源边缘代理（并且它们还有各自的活跃社区），因此开发人员不必编写和维护自己的边缘代理服务器。

最受欢迎的开源边缘代理服务器之一被称为 Zuul，它是由 Netflix 构建的。Zuul 是一项边缘服务，可提供动态路由、监控、弹性和安全性等。Zuul 被打包为 Java 库。用 Java 框架 Spring Boot 编写的服务可以使用嵌入式 Zuul 服务来提供边缘代理功能。在本秘笈中，我们将详细介绍如何构建一个小型 Zuul 边缘代理，并将其配置为将请求路由到我们的服务。

2.2.1 操作说明

继续沿用我们在第 1 章中提出的示例应用程序，假设我们的图像消息应用程序（从现在开始将其称为 pichat）最初是作为 Ruby on Rails 单体架构应用程序代码库实现的。当产品首次启动时，我们将该应用程序部署到单个 Elastic Load Balancer（ELB）后的 Amazon Web Services 中。我们为该单体架构应用程序创建了一个单一的自动缩放组（Auto Scale Group，ASG），称为 pichat-asg。

该自动缩放组（ASG）中的每个 EC2 实例都在运行 NGINX。NGINX（也称为 EngineX）是一个很强大的高性能 Web 和反向代理服务，也提供 IMAP/POP3/SMTP 服务，因此，它既可处理对静态文件（如图像、JavaScript、CSS）的请求，也可以代理单体架构应用程序的请求（该单体架构应用程序运行在为 Rails 应用程序提供服务的同一主机上）。SSL 在 ELB 终止，而 HTTP 请求则转发到 NGINX 中。可通过虚拟私有云（Virtual Private Cloud，VPC）中的 DNS monolith.pichat-int.me 名称访问 ELB。

现在，我们已经创建了单一的附件服务 attachment-service，该服务可以处理通过平台发送的消息中附加的视频和图像。这个 attachment-service 是用 Java 编写的，使用 Spring Boot 平台并部署在它自己的 ASG 中（称为 attachment-service-asg），后者具有自己的 ELB。

我们已经创建了一个私有 DNS 记录，称为 attachment-service.pichat-int.me，它就指向此 ELB。

在考虑到这种架构和拓扑之后，我们现在希望根据路径将请求从公共 Internet 中路由到我们的 Rails 应用程序或新创建的附件服务。

2.2.2 实战操作

本秘笈需要执行以下操作。

（1）为了演示如何使用 Zuul 将请求路由到服务，首先需要创建一个基本的 Java 应用程序，它将被用作我们的边缘代理服务。Java 项目 Spring Cloud 提供了一个嵌入式 Zuul 服务，使得通过 zuul 库创建服务非常简单。我们将从创建一个基本的 Java 应用程序开始。据此，创建包含以下内容的 build.gradle 文件：

```
group 'com.packtpub.microservices'
version '1.0-SNAPSHOT'

buildscript {
    repositories {
        mavenCentral()
    }
    dependencies {
        classpath "org.springframework.boot:spring-boot-gradle-plugin:1.4.4.RELEASE"
        classpath "io.spring.gradle:dependency-management-plugin:0.5.6.RELEASE"
    }
}

apply plugin: 'java'
apply plugin: 'org.springframework.boot'
apply plugin: 'io.spring.dependency-management'

sourceCompatibility = 1.8

repositories {
    mavenCentral()
}

dependencyManagement {
    imports {
```

```
        mavenBom 'org.springframework.cloud:spring-cloud-
netflix:1.4.4.RELEASE'
    }
}

dependencies {
    compile group: 'org.springframework.boot', name: 'spring-boot-
starter-web', version: '1.4.4.RELEASE'
    compile group: 'org.springframework.cloud', name: 'spring-
cloud-starter-zuul'
    testCompile group: 'junit', name: 'junit', version: '4.12'
}
```

（2）创建一个名为 EdgeProxyApplication 的类。这将作为我们的应用程序的入口点，具体如下：

```
package com.packtpub.microservices.ch02.edgeproxy;

import org.springframework.boot.SpringApplication;
import org.springframework.boot.autoconfigure.SpringBootApplication;
import org.springframework.cloud.netflix.zuul.EnableZuulProxy;

@EnableZuulProxy
@SpringBootApplication
public class EdgeProxyApplication {
    public static void main(String[] args) {
        SpringApplication.run(EdgeProxyApplication.class, args);
    }
}
```

（3）在应用程序的 src/main/resources 目录中创建一个名为 application.yml 的文件。该文件将指定你的路由配置。在本示例中，我们假设可以在 monolith.pichat-int.me 内部主机上访问我们的单体架构应用程序，并且希望将/signup 和/auth/login 路径公开到公共 Internet 中，具体如下：

```
zuul:
    routes:
        signup:
            path: /signup
            url: http://monolith.pichat-int.me
        auth:
            path: /auth/login
            url: http://monolith.pichat-int.me
```

（4）使用./gradlew bootRun 启动项目，你应该可以访问/signup 和/auth/login 两个 URL，并且这两个 URL 将被代理到单体架构应用程序中。

（5）我们还需要将附件服务 attachment-service 的 URL 也公开到 Internet 上。该附件服务将公开以下端点：

```
POST /                    # 创建附件
GET /                     # 提取附件，可以通过 message_id 过滤
DELETE /:attachment_id    # 删除指定的附件
GET /:id                  # 获取指定的附件
```

（6）我们需要决定在公共 API 中使用哪些路径。可修改 application.properties 以添加以下条目：

```yaml
zuul:
    routes:
        signup:
            path: /signup
            url: http://monolith.pichat-int.me
        auth:
            path: /auth/login
            url: http://monolith.pichat-int.me
        attachments:
            path: /attachments/**
            url: http://attachment-service.pichat-int.me
```

（7）现在，所有对/attachments/*的请求都将转发到 attachment-service 和 signup 中，并且 auth/login 将继续由单体架构应用程序提供服务。

（8）我们可以通过在本地运行服务并将请求发送到 localhost:8080/signup、localhost:8080/auth/login 和 localhost:8080/attachments/foo 来进行测试。你应该能够看到，请求已被路由到相应的服务。当然，该服务将以出错消息进行响应，这是因为无法解析 attachment-service.pichat-int.me，但这恰好表明路由正在按预期工作，具体如下：

```
$ curl -D - http://localhost:8080/attachments/foo
HTTP/1.1 500
X-Application-Context: application
Content-Type: application/json;charset=UTF-8
Transfer-Encoding: chunked
Date: Tue, 27 Mar 2018 12:52:21 GMT
Connection: close

{"timestamp":1522155141889,"status":500,"error":"Internal Server
```

```
Error","exception":"com.netflix.zuul.exception.ZuulException","mess
age":"attachment-service.pichat-int.me"}%
```

2.3 通过边车模式扩展服务

当你开始开发微服务时，通常会在每个服务中嵌入一定数量的样板（Boilerplate）。在服务之间经常会复制日志记录、度量（Metric）和配置等，从而导致大量样板以及复制和粘贴的代码。随着架构的增长和开发的服务的增加，这种设置变得越来越难维护。通常的结果是，你最终会获得多种不同的日志记录、度量、服务发现等方式，从而导致系统难以调试和维护。即使是更改一些非常简单的东西（如度量名称空间）或向服务发现客户端添加功能这样简单的操作，也可能需要多个团队和代码库的协调。更严重的是，由于日志记录、度量和服务发现约定不一致，你的微服务架构将继续增长，从而使开发人员更难以操作，最终导致整体操作十分困难。

2.3.1 理论阐释

所谓"边车"其实就是摩托车的挎斗。在现实生活中，人们会通过在一辆两轮摩托车上加装一个挎斗来组成一辆三轮摩托车，从而拓展其装载能力，这就是在不改变其原有功能的基础上增加新的服务。边车模式（Sidecar Pattern）就是指通过在同一台计算机上运行的单独进程或容器来扩展服务功能的模式。上面我们介绍的常用功能（如度量、日志记录、服务发现、配置甚至是网络 RPC 等）都可以从应用程序中被分解出来，由与程序一起运行的边车服务进行处理。通过在单独的进程中实现这些常用功能（所有服务都可以使用这个单独的进程），开发人员就可以在架构中轻松实现这些共享功能的标准化，这就是边车模式带来的好处。

实现边车模式的常见方法是构建一个很小的独立进程，该进程通过常用协议（如 HTTP）公开某些功能。例如，想象一下，你希望所有服务都使用集中式服务发现（Service-Discovery）服务，而不是依赖在每个应用程序的配置中设置 DNS 主机和端口。要实现这一目标，你需要为服务发现服务提供最新的客户端库，并且这些库需要使用你编写服务和单体架构应用程序时用到的所有语言提供。一种更好的方法是与每个运行服务发现客户端的服务并行运行一个边车进程，这样你的服务就可以将请求代理到边车进程，并让它确定将请求发送到哪里。边车模式的另一个好处是，开发人员可以将边车配置为围绕服务之间发出的网络 RPC 请求发出一致的度量。

边车模式是一种很常见的模式，因此也有多种开源解决方案可用。在本秘笈中，我们将使用 spring-cloud-netflix-sidecar，这是一个包含简单 HTTP API 的项目，该 API 允许非 Java 虚拟机（Java Virtual Machine，JVM）应用程序使用 JVM 客户端库。该 Netflix 边车假定你使用的是 Eureka（这是一个旨在支持客户端的服务发现需求的服务注册表）。我们将在后续章节中更详细地讨论服务发现。该边车还假定你的非 JVM 应用程序正在提供运行状况检查端点，并将使用此端点向 Eureka 通告其运行状况。我们的 Rails 应用程序公开了像 /health 这样的端点。当正常运行时，它将返回带有键状态和 UP 值的很小的 JSON 有效负载。

2.3.2　实战操作

本秘笈需要执行以下操作。

（1）创建一个基本的 Spring Boot 服务。它包括 Spring Boot Gradle 插件，并将为 Spring Boot 和 spring-cloud-netflix-sidecar 项目添加依赖项，具体如下：

```
group 'com.packtpub.microservices'
version '1.0-SNAPSHOT'

buildscript {
    repositories {
        mavenCentral()
    }
    dependencies {
        classpath "org.springframework.boot:spring-boot-gradle-plugin:1.4.4.RELEASE"
        classpath "io.spring.gradle:dependency-management-plugin:0.5.6.RELEASE"
    }
}

apply plugin: 'java'
apply plugin: 'org.springframework.boot'
apply plugin: 'io.spring.dependency-management'

sourceCompatibility = 1.8

repositories {
    mavenCentral()
}
```

```
dependencyManagement {
    imports {
        mavenBom 'org.springframework.cloud:spring-cloud-netflix:1.4.4.RELEASE'
    }
}

dependencies {
    compile group: 'org.springframework.boot', name: 'spring-boot-starter-web', version: '1.4.4.RELEASE'
    compile group: 'org.springframework.cloud', name: 'spring-cloud-netflix-sidecar', version: '1.4.4.RELEASE'
    testCompile group: 'junit', name: 'junit', version: '4.12'
}
```

（2）现在我们已经做好了准备，可以创建一个简单的 Spring Boot 应用程序。我们将使用@EnableSidecar 注解，其中还包括@EnableZuulProxy、@EnableCircuitBreaker 和 @EnableDiscoveryClient 注解，具体如下：

```
package com.packtpub.microservices;

import org.springframework.boot.SpringApplication;
import org.springframework.boot.autoconfigure.EnableAutoConfiguration;
import org.springframework.cloud.netflix.sidecar.EnableSidecar;
import org.springframework.stereotype.Controller;

@EnableSidecar
@Controller
@EnableAutoConfiguration
public class SidecarController {
    public static void main(String[] args) {
        SpringApplication.run(SidecarController.class, args);
    }
}
```

（3）Netflix 边车应用程序需要一些配置设置，因此可以创建一个名为 application.yml 的新文件，其内容如下：

```
server:
    port: 5678

sidecar:
```

```
port: 3000
health-uri: http://localhost:3000/health
```

(4)边车现在将公开一个 API,该 API 允许非 JVM 应用程序查找在 Eureka 中注册的服务。如果我们的附件服务 attachment-service 已在 Eureka 注册,则边车会将 http://localhost:5678/attachment/1234 请求代理到 http://attachment-service.pichat-int.me/1234。

2.4 使用 API 网关将请求路由到服务

正如我们在其他秘笈中所看到的那样,微服务应提供特定的业务功能,并应围绕一个或多个领域概念(由有界上下文包围)进行设计。这种设计服务边界的方法可以很好地指导你实现简单、可独立伸缩的服务,这些服务可以由专门针对你的应用程序或业务的特定领域的单个团队来管理和部署。

在设计用户界面时,客户端通常会聚合来自各种后端微服务的相关但截然不同的实体。例如,在我们虚构的图像消息应用程序中,显示实际消息的屏幕可能包含来自消息服务、媒体服务、喜欢服务、评论服务等的信息。所有这些信息可能收集起来都很烦琐,并且可能导致向后端的大量往返请求。

例如,将 Web 应用程序从使用服务器端渲染 HTML 的单体架构应用程序移植到如图 2-1 所示的单页 JavaScript 应用程序中,可以很容易导致单个页面加载数百个 XMLHttpRequest。

为了减少对后端服务的往返请求,可以考虑创建一个或多个 API 网关,以提供可满足客户端需求的 API。API 网

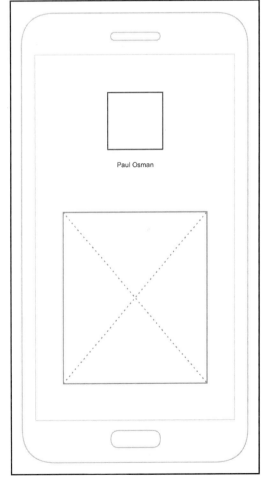

图 2-1

关可用于呈现后端实体的单一视图，从而使得连接该 API 的客户端更容易获取数据。例如，在上述我们假想的示例中，对单个消息端点的请求就可能返回有关消息本身、消息中包含的媒体、喜欢和评论，以及其他信息的所有信息。

可以使用扇出（Fan-Out）请求模式从各种后端服务同时收集这些实体，如图 2-2 所示。

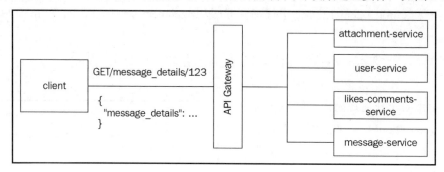

图 2-2

原　　文	译　　文
client	客户端
API Gateway	API 网关

2.4.1　设计时需要考虑的问题

使用 API 网关提供对微服务的访问的好处之一是，你可以为特定客户端创建一个单一的、一致性的 API。在大多数情况下，你需要为移动客户端创建一个特定的 API，甚至可能为 iOS 创建一个 API，为 Android 创建一个 API。API 网关的这种实现通常称为服务于前端的后端（Backend For Frontend，BFF），因为它为每个前端应用程序提供了一个逻辑后端。Web 应用程序与移动设备客户端的需求截然不同。

在本示例中，我们将专注于创建一个端点，该端点提供消息视图屏幕所需的所有数据。这包括消息本身以及附件、发送者的用户详细信息，以及其他的消息接收者。如果该消息是公开的，则还可以包含喜欢和评论（假设它们是由单独的服务提供的）。我们的端点可能看起来如下所示：

```
GET /message_details/:message_id
```

该端点将返回类似于以下内容的响应：

```
{
    "message_details": {
```

```
    "message": {
        "id": 1234,
        "body": "Hi There!",
        "from_user_id": "user:4321"
    },
    "attachments": [{
        "id": 4543,
        "media_type": 1,
        "url": "http://..."
    }],
    "from_user": {
        "username": "paulosman",
        "profile_pic": "http://...",
        "display_name": "Paul Osman"
    },
    "recipients": [
        ...
    ],
    "likes": 200,
    "comments": [{
        "id": 943,
        "body": "cool pic",
        "user": {
            "username": "somebody",
            "profile_pic": "http://..."
        }
    }]
}
```

此响应包含客户端在显示消息视图屏幕时所需的一切。数据本身来自各种服务，但是，正如我们将看到的那样，API 网关执行了这些麻烦的工作来发出请求并汇总响应。

2.4.2 实战操作

API 网关负责公开 API，执行多次服务调用，汇总结果并将其返回给客户端。Finagle Scala 框架将这些服务调用表示为 Future（这也表示了其依赖关系）。为了与本书中的其他示例保持一致，我们将使用 Spring Boot 框架在 Java 中构建一个小型示例网关服务。具体操作如下。

（1）创建项目的大致结构。我们可以新建一个 Java 项目，并将以下依赖项和插件添加到 Gradle 构建文件中。在此秘笈中，我们将使用 Spring Boot 和 Hystrix，具体如下：

```
plugins {
    id 'org.springframework.boot' version '1.5.9.RELEASE'
}

group 'com.packtpub.microservices'
version '1.0-SNAPSHOT'

apply plugin: 'java'

sourceCompatibility = 1.8

repositories {
    mavenCentral()
}

dependencies {
    compile group: 'org.springframework.boot', name: 'spring-boot-starter-web', version: '1.5.9.RELEASE'
    compile group: 'com.netflix.hystrix', name: 'hystrix-core', version: '1.0.2'
    testCompile group: 'junit', name: 'junit', version: '4.12'
}
```

查看 2.4.1 节中的 JSON 示例，很明显，我们正在收集和聚合一些截然不同的领域的数据。就本示例而言，我们将假设有一个消息服务，该服务检索有关消息的信息（包括喜欢、评论和附件）以及一个用户服务。网关服务将调用消息服务以检索消息本身，然后调用其他服务以获取关联的数据，我们将在单个的响应中将它们拼接在一起。就本秘笈而言，将假设消息服务在端口 4567 上运行，而用户服务在端口 4568 上运行。我们将创建一些桩代码服务来模拟这些假设的微服务的数据。

（2）创建一个表示 Message 数据的模块，具体如下：

```
package com.packtpub.microservices.gateway.models;

import com.fasterxml.jackson.annotation.JsonIgnoreProperties;
import com.fasterxml.jackson.annotation.JsonProperty;

@JsonIgnoreProperties(ignoreUnknown = false)
public class Message {

    private String id;
```

```
    private String body;

    @JsonProperty("from_user_id")
    private String fromUserId;

    public String getId() {
        return id;
    }

    public void setId(String id) {
        this.id = id;
    }

    public String getBody() {
        return body;
    }

    public void setBody(String body) {
        this.body = body;
    }

    public String getFromUserId() {
        return fromUserId;
    }

    public void setFromUserId(String fromUserId) {
        this.fromUserId = fromUserId;
    }
}
```

非依赖性服务调用以非阻塞的异步方式进行很重要。幸运的是，Hystrix 可以选择异步执行命令，并返回 Future<T>。

（3）使用以下类创建一个新程序包，如 com.packtpub.microservices.gateway.commands。

❑ 创建包含以下内容的 AttachmentCommand 类：

```
package com.packtpub.microservices.gateway.commands;

import com.netflix.hystrix.HystrixCommand;
import com.netflix.hystrix.HystrixCommandGroupKey;
import org.springframework.http.ResponseEntity;
import org.springframework.web.client.RestTemplate;
```

```java
public class AttachmentCommand extends HystrixCommand<String> {
    private String messageId;

    public AttachmentCommand(String messageId) {
super(HystrixCommandGroupKey.Factory.asKey("AttachmentCommand"));
        this.messageId = messageId;
    }

    @Override
    public String run() {
        RestTemplate template = new RestTemplate();
        String attachmentsUrl = "http://localhost:4567/message/" +
messageId + "/attachments";
        ResponseEntity<String> response =
template.getForEntity(attachmentsUrl, String.class);
        return response.getBody();
    }
}
```

- 创建包含以下内容的 CommentCommand 类：

```java
package com.packtpub.microservices.commands;

import com.netflix.hystrix.HystrixCommand;
import com.netflix.hystrix.HystrixCommandGroupKey;
import org.springframework.http.ResponseEntity;
import org.springframework.web.client.RestTemplate;

public class CommentCommand extends HystrixCommand<String> {

    private String messageId;

    public CommentCommand(String messageId) {
super(HystrixCommandGroupKey.Factory.asKey("CommentGroup"));
        this.messageId = messageId;
    }

    @Override
    public String run() {
        RestTemplate template = new RestTemplate();
        String commentsUrl = "http://localhost:4567/message/" +
messageId + "/comments";
        ResponseEntity<String> response =
```

```
template.getForEntity(commentsUrl, String.class);
    return response.getBody();
    }
}
```

- 创建包含以下内容的 LikeCommand 类：

```
package com.packtpub.microservices.commands;

import com.netflix.hystrix.HystrixCommand;
import com.netflix.hystrix.HystrixCommandGroupKey;
import org.springframework.http.ResponseEntity;
import org.springframework.web.client.RestTemplate;

public class LikeCommand extends HystrixCommand<String> {

    private String messageId;

    public LikeCommand(String messageId) {
        super(HystrixCommandGroupKey.Factory.asKey("Likegroup"));
        this.messageId = messageId;
    }

    @Override
    public String run() {
        RestTemplate template = new RestTemplate();
        String likesUrl = "http://localhost:4567/message/" +
messageId + "/likes";
        ResponseEntity<String> response =
template.getForEntity(likesUrl, String.class);
        return response.getBody();
    }
}
```

- 我们的 MessageClient 类与前述示例有些不同，它不是从服务响应中返回 JSON 字符串，而是返回对象的表示形式。在本示例中，它就是 Message 类的实例，具体如下：

```
package com.packtpub.microservices.commands;

import com.netflix.hystrix.HystrixCommand;
import com.netflix.hystrix.HystrixCommandGroupKey;
import com.packtpub.microservices.models.Message;
```

```
import org.springframework.web.client.RestTemplate;

public class MessageClient extends HystrixCommand<Message> {

    private final String id;
    public MessageClient(String id) {
super(HystrixCommandGroupKey.Factory.asKey("MessageGroup"));
        this.id = id;
    }

    @Override
    public Message run() {
        RestTemplate template = new RestTemplate();
        String messageServiceUrl = "http://localhost:4567/message/" + id;
        Message message = template.getForObject(messageServiceUrl, Message.class);
        return message;
    }
}
```

- 创建包含以下内容的 UserCommand 类:

```
package com.packtpub.microservices.commands;

import com.netflix.hystrix.HystrixCommand;
import com.netflix.hystrix.HystrixCommandGroupKey;
import org.springframework.http.ResponseEntity;
import org.springframework.web.client.RestTemplate;

public class UserCommand extends HystrixCommand<String> {

    private String id;

    public UserCommand(String id) {
        super(HystrixCommandGroupKey.Factory.asKey("UserGroup"));
        this.id = id;
    }

    @Override
    public String run() {
        RestTemplate template = new RestTemplate();
        String userServiceUrl = "http://localhost:4568/user/" + id;
```

```
        ResponseEntity<String> response =
template.getForEntity(userServiceUrl, String.class);
        return response.getBody();
    }
}
```

（4）在单个控制器中将这些 Hystrix 命令的执行组合在一起，该控制器将我们的 API 公开为/message_details/:message_id 端点，具体如下：

```
package com.packtpub.microservices;

import com.fasterxml.jackson.databind.ObjectMapper;
import com.packtpub.microservices.commands.*;
import com.packtpub.microservices.models.Message;
import org.springframework.boot.SpringApplication;
import org.springframework.http.MediaType;
import org.springframework.boot.autoconfigure.SpringBootApplication;
import org.springframework.web.bind.annotation.PathVariable;
import org.springframework.web.bind.annotation.RequestMapping;
import org.springframework.web.bind.annotation.RestController;
import java.io.IOException;
import java.io.StringWriter;
import java.util.HashMap;
import java.util.Map;
import java.util.concurrent.ExecutionException;
import java.util.concurrent.Future;

@SpringBootApplication
@RestController
public class MainController {

    @RequestMapping(value = "/message_details/{id}", produces = MediaType.APPLICATION_JSON_UTF8_VALUE)
    public Map<String, HashMap<String, String>> messageDetails(@PathVariable String id)
            throws ExecutionException, InterruptedException, IOException {

        Map<String, HashMap<String, String>> result = new HashMap<>();
        HashMap<String, String> innerResult = new HashMap<>();

        Message message = new MessageClient(id).run();
```

```
            String messageId = message.getId();

            Future<String> user = new
UserClient(message.getFromUserId()).queue();
            Future<String> attachments = new
AttachmentClient(messageId).queue();
            Future<String> likes = new LikeClient(messageId).queue();
            Future<String> comments = new
CommentClient(messageId).queue();

            ObjectMapper mapper = new ObjectMapper();
            StringWriter writer = new StringWriter();
            mapper.writeValue(writer, message);

            innerResult.put("message", writer.toString());
            innerResult.put("from_user", user.get());
            innerResult.put("attachments", attachments.get());
            innerResult.put("comments", comments.get());
            innerResult.put("likes", likes.get());

            result.put("message_details", innerResult);

            return result;
        }

    public static void main(String[] args) {
        SpringApplication.run(MainController.class, args);
        }
}
```

（5）现在可以使用./gradlew bootRun 运行该服务，并通过以下请求进行测试：

```
$ curl -H "Content-Type: application/json"
http://localhost:8080/message_details/1234
```

2.5 使用 Hystrix 停止级联故障

复杂系统中的故障可能很难诊断。在很多情况下，症状可能出现在远离病因的地方。由于某些下游服务会管理个人资料图片或与用户个人资料有关系的某些服务，因此用户在登录期间可能会开始遇到高于正常的错误率。一种服务中的错误通常会不必要地传播到用户请求,并且对用户体验产生不利影响,因此也阻碍了他们对应用程序的信任。此外,服务失败也会产生级联效应,将很小的系统中断变成严重的、影响客户的事件。也就是说,当各项服务的功能紧密耦合时,很容易出现"雪崩效应"。因此,在设计微服务时,考虑故障隔离并决定如何处理不同的故障情况非常重要。

2.5.1 理论阐释

可以使用许多模式来提高分布式系统的弹性。例如,断路器(Circuit Breaker)就是一种常见的模式,如果对某一个微服务的请求出现了大量超时的情况,则让这个微服务的对外通信中断,实现快速失败。所以,断路器的作用就像电路中的保险丝一样。

断路器还可以对已经被断路的微服务进行检测,查看微服务是否已经恢复正常,如果是,则恢复该应用并对外提供服务。Michael Nygard 编写的图书 *Release It!*(释放它!)首次描述了断路器的工作原理。正常情况下,微服务的断路器处于闭合(Closed)状态,这意味着所有请求都将被正常发送到下游服务。但是,如果断路器在短时期内太快接收到太多的故障(访问失败率或故障率到达一定阈值),则断路器的状态更改为打开(Open),并启动快速故障处理,对该微服务的请求全部被中断。断路器不会等待下游服务再次失败并增加失败服务的负载,而是向上游服务发送一条错误消息,从而使不堪重负的服务时间得以恢复。经过一段时间之后,断路器将允许一个请求访问微服务,并检测是否调用成功。如果成功,则闭合断路器;如果失败,则断路器保持打开的状态。

有许多框架和库都可以实现断路器。诸如 Twitter 的 Finagle 之类的某些框架会自动将每个 RPC 调用都包装在断路器中。在本示例中,我们将使用流行的 Netflix 库 hystrix。Hystrix 是一个通用的容错库,它将隔离的代码构造为命令。当执行命令时,它将检查断路器的状态,以决定是发出请求还是使请求短路。

2.5.2 实战操作

Hystrix 可作为 Java 库使用,因此我们将通过构建一个小型 Java Spring Boot 应用程序来演示其用法。具体操作如下。

(1)创建一个新的 Java 应用程序,并将依赖项添加到 build.gradle 文件中,具体如下:

```
plugins {
    id 'org.springframework.boot' version '1.5.9.RELEASE'
}
```

```groovy
group 'com.packetpub.microservices'
version '1.0-SNAPSHOT'

apply plugin: 'java'

sourceCompatibility = 1.8

repositories {
    mavenCentral()
}

dependencies {
    compile group: 'org.springframework.boot', name: 'spring-boot-starter-web', version: '1.5.9.RELEASE'
    compile group: 'com.netflix.hystrix', name: 'hystrix-core', version: '1.0.2'
    testCompile group: 'junit', name: 'junit', version: '4.12'
}
```

（2）为简单起见，我们将创建一个仅返回一条消息的简易 MainController。这虽然只是一个简易示例，但是它同样演示了上游服务进行的下游调用。首先，我们的应用程序将仅返回硬编码的"Hello,World!"信息。接下来，我们将字符串移出至 Hystrix 命令中。最后，我们将该消息移动到包装在 Hystrix 命令内的服务调用中：

```java
package com.packtpub.microservices;

import org.springframework.boot.SpringApplication;
import org.springframework.boot.autoconfigure.EnableAutoConfiguration;
import org.springframework.boot.autoconfigure.SpringBootApplication;
import org.springframework.web.bind.annotation.RequestMapping;
import org.springframework.web.bind.annotation.RestController;

@SpringBootApplication
@EnableAutoConfiguration
@RestController
public class MainController {
    @RequestMapping("/message")
    public String message() {
        return "Hello, World!";
    }

    public static void main(String[] args) {
```

```
        SpringApplication.run(MainController.class, args);
    }
}
```

(3)将消息移出至 HystrixCommand 中:

```
package com.packtpub.microservices;

import com.netflix.hystrix.HystrixCommand;
import com.netflix.hystrix.HystrixCommandGroupKey;

public class CommandHelloWorld extends HystrixCommand<String> {

    private String name;

    CommandHelloWorld(String name) {
super(HystrixCommandGroupKey.Factory.asKey("ExampleGroup"));
        this.name = name;
    }

    @Override
    public String run() {
        return "Hello, " + name + "!";
    }
}
```

(4)替换 MainController 中的方法以使用 HystrixCommand:

```
@RequestMapping("/message")
public String message() {
    return new CommandHelloWorld("Paul").execute();
}
```

(5)将消息生成移至另一个服务中。注意:我们在此处对假设的消息服务 URL 进行了硬编码,这并不是一个好习惯,仅出于演示目的。具体如下:

```
package com.packtpub.microservices;

import com.netflix.hystrix.HystrixCommand;
import com.netflix.hystrix.HystrixCommandGroupKey;
import org.springframework.http.ResponseEntity;
import org.springframework.web.client.RestTemplate;

public class CommandHelloWorld extends HystrixCommand<String> {
```

```
    CommandHelloWorld() {
super(HystrixCommandGroupKey.Factory.asKey("ExampleGroup"));
    }

    @Override
    public String run() {
        RestTemplate restTemplate = new RestTemplate();
        String messageResourceUrl = "http://localhost:4567/";
        ResponseEntity<String> response =
restTemplate.getForEntity(messageResourceUrl, String.class);
        return response.getBody();
    }

    @Override
    public String getFallback() {
        return "Hello, Fallback Message";
    }
}
```

（6）更新 MainController 类以包含以下内容：

```
package com.packetpub.microservices;

import org.springframework.boot.SpringApplication;
import org.springframework.boot.autoconfigure.EnableAutoConfiguration;
import org.springframework.boot.autoconfigure.SpringBootApplication;
import org.springframework.web.bind.annotation.RequestMapping;
import org.springframework.web.bind.annotation.RestController;

@SpringBootApplication
@EnableAutoConfiguration
@RestController
public class MainController {

    @RequestMapping("/message")
    public String message() {
        return new CommandHelloWorld().execute();
    }

    public static void main(String[] args) {
        SpringApplication.run(MainController.class, args);
    }
}
```

（7）现在，我们的 MainController 类进行了服务调用，并包装在 Hystrix 命令中，以生成一条消息发送回客户端。要测试它是否能够正常工作，可以创建一个非常简单的服务来生成消息字符串。sinatra 是一个易于使用的 Ruby 库，非常适合创建测试服务。

现在创建一个名为 message-service.rb 的新文件，具体如下：

```
require 'sinatra'

get '/' do
    "Hello from Sinatra"
end
```

（8）要运行该服务，可以运行 ruby message-service.rb，然后向启用了 Hystrix 的服务发出一些示例请求。可以通过修改服务返回 503 来模拟失败，这表明该服务暂时不堪重负，具体如下：

```
require 'sinatra'

get '/' do
    halt 503, 'Busy'
end
```

你的 Spring 服务现在应该会尝试访问服务，但是当遇到 503 时，使用备用（Fallback）处理流程中的值。此外，经过多次这样的尝试之后，命令的断路器将打开，并且该服务将在一段时间内默认为使用备用处理流程。

2.6 速率限制

除了诸如断路之类的技术外，也可以使用速率限制方法有效防止分布式系统中的级联故障。速率限制可以有效地防止垃圾邮件，防御拒绝服务（Denial of Service，DoS）攻击，并防止系统的各个部分因过多的同时请求而变得过载。速率限制通常实现为全局限制或每个客户端限制，一般来说是代理或负载均衡器的一部分。在本秘笈中，我们将使用前文介绍过的 NGINX，这是一种流行的开源负载均衡器、Web 服务器和反向代理。

2.6.1 理论阐释

大多数速率限制实现都使用漏桶算法（Leaky-Bucket Algorithm），该算法起源于计算机网络交换机和电信网络。顾名思义，漏桶算法是基于一个桶的隐喻，即水（请求）

先进入一个会漏的桶中，漏桶以一定的速度出水，当水流入的速度过大时则会直接溢出，通过这种方式，漏桶算法可以控制一个恒定的速率。外面的水可能会突然注入桶中，但漏桶中存在的水始终是稳定的，其出水的速率也是固定的。如果外面注水的速度快于漏桶出水的速度，则最终会导致漏桶溢出。在这种情况下，溢出表示将会有请求被丢弃。

开发人员可以实现自己的速率限制解决方案，甚至还有其他的开放源代码也可以用于该算法的实现。但是，使用诸如 NGINX 之类的产品进行速率限制要容易得多。因此，本秘笈将配置 NGINX 以代理对微服务的请求。

2.6.2 实战操作

本秘笈需要执行以下操作。

（1）通过运行以下命令来安装 NGINX：

```
apt-get install nginx
```

（2）NGINX 有一个配置文件 nginx.conf。在基于 Ubuntu 的 Linux 系统上，该目录可能位于/etc/nginx/nginx.conf 目录中。请打开该文件并查找 http 块以添加如下内容：

```
limit_req_zone $binary_remote_addr zone=mylimit:10m rate=10r/s;
server {
    location /auth/signin {
        limit_req zone=mylimit;
        proxy_pass http://my_upstream;
    }
}
```

从上述代码中可以看到，速率限制是通过两个配置指令实现的。limit_req_zone 指令定义了用于速率限制的参数。在本示例中，我们基于客户端的 IP 地址实现了每秒 10 个请求的速率限制。limit_req 指令将该速率限制应用于特定的路径或位置。在本示例中，速率限制被应用于对/auth/signin 的所有请求，这是因为我们不希望机器人通过脚本来创建账户！

2.7 使用服务网格解决共同关注的问题

随着 Web 服务框架和标准的发展，样板或共享的应用程序问题的数量减少了。这是因为，我们总体上确认了应用程序的哪些部分是通用的，因此不需要每个程序员或团队都重新实现。当人们首次启动计算机联网时，编写支持网络的应用程序的程序员必须担心许多底层细节，但是这些细节现在已由操作系统的网络堆栈抽象出来了。同样，所有微服务都存在某些普遍关注的问题。于是，就有了 Twitter 的 Finagle 之类的框架将所有网络调用都包装在断路器中，从而提高了容错能力并隔离了系统中的故障。Finagle 和 Spring Boot 是我们在大多数秘笈中使用的 Java 框架，它们都支持公开标准度量端点，该端点将对为微服务收集的基本网络、JVM 和应用程序度量进行标准化。

2.7.1 理论阐释

每个微服务都应考虑许多共享的应用程序问题。从可观察性的角度来看，服务应努力发出一致的度量和结构化日志。为了提高系统的可靠性，服务应将网络调用包装在断路器中，并实现一致的重试和后退逻辑。为了支持网络和服务拓扑的更改，服务应考虑实现客户端负载均衡并使用集中式服务发现。

与其在每一项服务中实现所有功能，不如将它们抽象到应用程序代码之外，作为一个可以单独维护和操作的内容，这是一种更理想的做法。像操作系统网络堆栈的功能一样，如果这些功能中的每一个都以某种方式实现，并且这种方式对于我们的应用程序来说是完全可靠的，那么我们就不必担心它们是否可用。这就是服务网格（Service Mesh）背后的思路。

运行服务网格配置需要在网络代理后面运行系统中的每个微服务。这些服务不是直接相互通信，而是通过各自的代理进行通信，它们将作为边车安装。更直白一点说，你的服务将与在本地主机上运行的自己的代理进行通信。通过服务代理发送网络请求时，代理可以控制发出什么度量和输出什么日志消息。代理还可以直接与你的服务注册表集成，并在活动节点之间平均分配请求，跟踪故障并在达到特定阈值时选择快速故障处理。以这种配置运行系统可以减轻系统的操作复杂性，同时提高架构的可靠性和可观察性。

像本章中讨论的大多数秘笈一样，也有许多可用于运行服务网格的开源解决方案。这里我们将重点介绍 Linkerd，这是一个由 buoyant 构建和维护的开源代理服务器。Linkerd 的原作者在成立 buoyant 之前曾在 Twitter 工作，因此，Linkerd 吸收了 Twitter 团队所汲取的许多经验教训。它与 Finagle Scala 框架共享许多功能，但可以与以任何语言编写的服务一起使用。在本秘笈中，我们将逐步介绍如何安装和配置 Linkerd，并讨论如何使用它来控制 Ruby on Rails 单体架构 API 与我们新开发的媒体服务之间的通信。

2.7.2 实战操作

为了演示在代理后面运行服务，我们将安装并运行 Linkerd 实例，并将其配置为处理往返于你的服务的请求。Linkerd 网站上提供了如何在 Docker、Kubernetes 和其他选项中运行它的指导。为简单起见，我们将专注于在本地运行 Linkerd 和我们的服务。具体操作如下。

（1）从以下网址下载最新的 Linkerd 版本：

https://github.com/linkerd/linkerd/releases

（2）通过执行以下命令来提取压缩包：

```
$ tar xvfz linkerd-1.3.4.tgz
$ cd linkerd-1.3.4
```

（3）默认情况下，Linkerd 随附的配置将使用基于文件的服务发现。后面我们将会讨论这种方法的替代方法，但是当前我们将创建一个名为 disco/media-service 的新文件，它包含以下内容：

```
localhost 8080
```

（4）这会将主机名和端口映射到名为 media-service 的服务中。Linkerd 将使用此文件按名称查找服务，并确定主机名和端口映射。

（5）按以下方式运行 Linkerd：

```
$ ./linkerd-1.3.4-exec config/linkerd.yaml
```

（6）在端口 8080 上启动该服务。切换到 media-service 目录中并按以下方式运行该服务：

```
$ ./gradlew bootRun
```

（7）Linkerd 正在端口 4140 上运行。可使用以下请求测试该代理是否能正常工作：

```
$ curl -H "Host: attachment-service" http://localhost:4140/
```

第 3 章　服务间通信

本章包含以下操作秘笈。
- 从服务到服务的通信。
- 并发异步请求。
- 使用服务发现来查找服务。
- 服务器端负载均衡。
- 客户端负载均衡。
- 构建事件驱动的微服务。
- 不断演变的 API。

3.1　导　　语

在前述章节中，我们介绍了如何开始将单体架构应用程序代码库分解为微服务，以及将微服务公开到公共 Internet 中的最佳实践。到目前为止，我们假设所有微服务都是独立的应用程序，没有依赖关系。这些简单的微服务将接收请求、检索数据或写入数据库中，并将响应返回给客户端。这种线性工作流在实际系统中其实是很少见的。在现实世界的微服务架构中，服务经常需要调用其他服务才能满足用户的请求。典型的用户请求通常需要为系统中的服务创建数十个请求。

管理服务之间的通信也对开发人员提出了许多挑战。在服务可以与另一个服务进行对话之前，它将需要通过某种服务发现机制来定位该服务。在生成对下游服务的请求时，我们还需要一种在服务的各个实例之间分配流量的方法，该方法应该可以最大程度地减少延迟，并在不影响数据完整性的情况下平均分配负载。我们还需要考虑如何处理服务故障，并防止它们在整个系统中产生雪崩效应。

有时一个服务将需要与其他服务异步通信，在这种情况下，我们可以使用事件驱动的架构模式来创建响应式工作流。将我们的系统分解为多种服务还意味着不同的服务将独立地发展其 API，因此我们将需要一些方法来处理不会破坏上游服务的变更。

在本章中，我们将讨论旨在解决这些问题的秘笈。到本章结束时，你将能够自信地处理我们在微服务架构中必然需要的各种类型的交互。

3.2 从服务到服务的通信

在大型系统中,问题很少出现在服务本身中,而是更多地出现在服务之间的通信中。因此,我们需要仔细考虑从服务到服务通信中的各种挑战。

3.2.1 理论阐释

在讨论服务到服务的通信时,对系统中的信息流进行可视化很有用。数据的流动是双向的:一方面,数据将从客户端(上游)流动到数据库中;另一方面,以请求形式的事件总线(下游)也将以响应的形式返回。当我们提到上游服务时,一般指的是信息流中与用户更接近的系统组件;而当我们提到下游服务时,一般指的是远离用户的系统组件。换句话说,用户发出的请求被路由到服务,然后该服务再向其他下游服务发出请求,如图 3-1 所示。

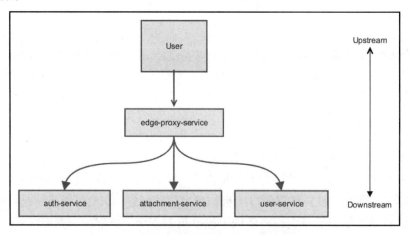

图 3-1

原　　文	译　　文
User	用户
Upstream	上游
Downstream	下游

在图 3-1 中,用户位于 edge-proxy-service(边缘代理服务)的上游,而边缘代理服务则位于 auth-service(验证服务)、attachment-service(附件服务)和 user-service(用户服

务）等服务的上游。

为了演示服务之间的通信，我们将创建一个简单的服务，该服务使用 Spring Boot Java 框架同步调用另一个服务。结合前面我们虚构的图像消息应用程序的示例，我们将创建一个 message-service 服务，负责发送消息。该 message-service 服务必须调用 social-service 社交图服务，以便在允许发送消息之前确定消息的发送者和接收者是否为好友。图 3-2 简要说明了这些服务之间的关系。

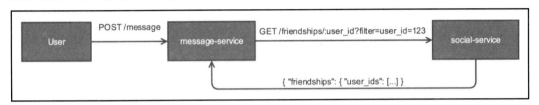

图 3-2

原　　文	译　　文
User	用户

可以看到，POST 请求是从用户发送到/message 端点的，它将被路由到 message-service。然后，message-service 服务使用/friendships/:user_id 端点向 social-service 服务发出 HTTP GET 请求。social-service 服务返回用户的好友关系的 JSON 表示。

3.2.2 实战操作

本秘笈需要执行以下操作。

（1）创建一个名为 message-service 的新 Java/Gradle 项目，并将以下内容添加到 build.gradle 文件中：

```
group 'com.packtpub.microservices'
version '1.0-SNAPSHOT'

buildscript {
    repositories {
        mavenCentral()
    }
    dependencies {
        classpath group: 'org.springframework.boot', name: 'spring-boot-gradle-plugin', version: '1.5.9.RELEASE'
    }
}
```

```
apply plugin: 'java'
apply plugin: 'org.springframework.boot'

sourceCompatibility = 1.8

repositories {
    mavenCentral()
}

dependencies {
    compile group: 'org.springframework.boot', name: 'spring-boot-starter-web'
    testCompile group: 'junit', name: 'junit', version: '4.12'
}
```

(2)创建一个名为 com.packtpub.microservices.ch03.message 的新程序包和一个名为 Application 的新类。这将是我们的服务的入口点。具体如下：

```
package com.packtpub.microservices.ch03.message;

import org.springframework.boot.SpringApplication;
import org.springframework.boot.autoconfigure.SpringBootApplication;

@SpringBootApplication
public class Application {
    public static void main(String[] args) {
        SpringApplication.run(Application.class, args);
    }
}
```

(3)创建模块。首先创建一个名为 com.packtpub.microservices.ch03.message.models 的程序包和一个名为 Message 的类。这是消息的内部表示。注意，这里我们为保持示例的简单起见，忽略了很多内容，并没有在此代码中真正持久化消息。具体如下：

```
package com.packtpub.microservices.ch03.message.models;

public class Message {

    private String toUser;
    private String fromUser;
    private String body;
```

```java
    public Message() {}

    public Message(String toUser, String fromUser, String body) {
        this.toUser = toUser;
        this.fromUser = fromUser;
        this.body = body;
    }

    public String getToUser() {
        return toUser;
    }

    public String getFromUser() {
        return fromUser;
    }

    public String getBody() {
        return body;
    }
}
```

(4)创建一个名为 com.packtpub.microservices.ch03.message.controllers 的新程序包和一个名为 MessageController 的新类。在目前这个阶段,控制器除了接收请求、解析 JSON 并返回消息实例外,并不需要做太多事情,这在以下代码中也可以看到:

```java
package com.packtpub.microservices.ch03.message.controllers;

import com.packtpub.microservices.models.Message;
import org.springframework.web.bind.annotation.*;

@RestController
public class MessageController {

    @RequestMapping(
            path="/messages",
            method=RequestMethod.POST,
            produces="application/json")
    public Message create(@RequestBody Message message) {
        return message;
    }
}
```

(5)要测试基本服务,我们可以运行它并尝试发送一个简单的请求,具体如下:

```
$ ./gradlew bootRun
Starting a Gradle Daemon, 1 busy Daemon could not be reused, use --
status for details

> Task :bootRun

  .   ____          _            __ _ _
 /\\ / ___'_ __ _ _(_)_ __  __ _ \ \ \ \
( ( )\___ | '_ | '_| | '_ \/ _` | \ \ \ \
 \\/  ___)| |_)| | | | | || (_| |  ) ) ) )
  '  |____| .__|_| |_|_| |_\__, | / / / /
 =========|_|==============|___/=/_/_/_/
 :: Spring Boot ::        (v1.5.9.RELEASE)

...
```

查看以下命令行:

```
$ curl -H "Content-Type: application/json" -X POST
http://localhost:8080/messages -d'{"toUser": "reader", "fromUser":
"paulosman", "body": "Hello, World"}'

{"toUser":"reader","fromUser":"paulosman","body":"Hello, World"}
```

现在我们有一个基本的服务正在运行,但是它很笨,做不了太多事情。本章不会涉及持久性保存的问题,但是我们可以通过与 social-service 服务进行通信来检索一些信息,以便在允许发送消息之前验证两个用户是否为好友。为演示起见,假设我们有一个可以正常工作的 social-service 服务,该服务可以让我们通过请求核对用户之间的关系,具体如下:

```
GET /friendships?username=paulosman&filter=reader

{
    "username": "paulosman",
    "friendships": [
        "reader"
    ]
}
```

(6) 在使用此服务之前,还需要创建一个模块来存储其响应。在 com.packtpub.microservices.ch03.message.models 程序包中,可以创建一个名为 UserFriendships 的类,其内容如下:

第3章 服务间通信

```java
package com.packtpub.microservices.ch03.message.models;

import com.fasterxml.jackson.annotation.JsonIgnoreProperties;

import java.util.List;

@JsonIgnoreProperties(ignoreUnknown = true)
public class UserFriendships {
    private String username;
    private List<String> friendships;

    public UserFriendships() {}

    public String getUsername() {
        return username;
    }

    public void setUsername(String username) {
        this.username = username;
    }

    public List<String> getFriendships() {
        return friendships;
    }

    public void setFriendships(List<String> friendships) {
        this.friendships = friendships;
    }
}
```

（7）修改 MessageController，添加一种方法来获取用户的好友关系列表，可以选择按用户名进行过滤。请注意，在此示例中，我们对 URL 进行的是硬编码，这是一种不好的做法。我们将在下一个秘笈中讨论替代方法。查看以下代码：

```java
private List<String> getFriendsForUser(String username, String filter) {
    String url = "http://localhost:4567/friendships?username=" + username + "&filter=" + filter;
    RestTemplate template = new RestTemplate();

    UserFriendships friendships = template.getForObject(url, UserFriendships.class);
    return friendships.getFriendships();
}
```

（8）修改之前编写的 create 方法。如果用户是好友，则将像以前一样继续并返回消息；如果用户不是好友，则该服务将返回 403 表示禁止该请求。代码如下：

```
@RequestMapping(
        path="/messages",
        method=RequestMethod.POST,
        produces="application/json")
public ResponseEntity<Message> create(@RequestBody Message message) {
    List<String> friendships = getFriendsForUser(message.getFromUser(), message.getToUser());

    if (friendships.isEmpty())
        return ResponseEntity.status(HttpStatus.FORBIDDEN).build();

    URI location = ServletUriComponentsBuilder
            .fromCurrentRequest().path("/{id}")
            .buildAndExpand(message.getFromUser()).toUri();
    return ResponseEntity.created(location).build();
}
```

3.3　并发异步请求

在先前的秘笈中，我们从消息服务到社交服务，对每个请求进行单个服务调用。这样的好处是实现起来非常简单，并且在使用诸如 Python、Ruby 或 JavaScript 之类的单线程语言时，通常是唯一的选择。当你对每个请求仅执行一次网络调用时，以这种方式同步执行网络调用是可以接受的——在这种情况下，调用阻塞了线程也没有关系，因为在完成调用之前你是无法响应用户的。但是，当你需要发出多个请求时，阻塞网络调用将严重影响应用程序的性能和可伸缩性。因此，我们需要一种简单的方法来利用 Java 的并发功能。

3.3.1　理论阐释

如果你使用 Scala 编写微服务，则可以考虑利用 Future 类型，该类型常用于表示异步计算。Finagle RPC 框架甚至使用 Future 作为其基础抽象之一，用于对依赖的 RPC 进行建模。Java 也有 Future，并且 Spring Boot 框架具有一些有用的实用程序，这些实用程序使打包网络调用变得非常容易，从而使它们变成异步方式并且不再阻塞。

在本秘笈中，我们将重新构建在先前秘笈中介绍过的消息服务。现在，我们将设想应用程序使用非对称关注模块，而不是检查消息的发送者和接收者是否为好友。为了使一个用户能够向另一个用户发送消息，这两个用户将必须互相关注。这要求 message-service 服务对 social-service 服务进行两次的网络调用，检查发送者是否关注了接收者，同时检查接收者是否也关注了发送者。图 3-3 简单表示了这两个服务之间的关系。

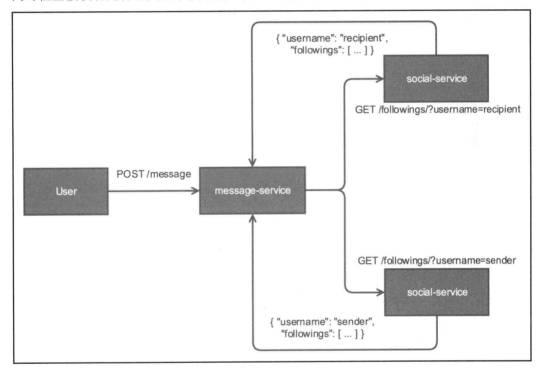

图 3-3

原　　文	译　　文
User	用户

Spring Boot 提供了很实用的工具，可通过 Java 的 CompletableFuture 类型使方法异步。我们将修改以前的 message-service 服务，以并发调用两个搜索服务。

3.3.2　实战操作

本秘笈需要执行以下操作。

（1）打开 MessageController 文件并插入以下内容：

```java
package com.packtpub.microservices.ch03.message.controllers;

import com.packtpub.microservices.models.Message;
import com.packtpub.microservices.models.UserFriendships;
import org.springframework.http.HttpStatus;
import org.springframework.http.ResponseEntity;
import org.springframework.web.bind.annotation.*;
import org.springframework.web.client.RestTemplate;
import org.springframework.web.servlet.support.ServletUriComponentsBuilder;

import java.net.URI;
import java.util.List;

@RestController
public class MessageController {

    @RequestMapping(
            path="/messages",
            method=RequestMethod.POST,
            produces="application/json")
    public ResponseEntity<Message> create(@RequestBody Message message) {
        List<String> friendships =
getFriendsForUser(message.getFromUser(), message.getToUser());

        if (friendships.isEmpty())
            return
ResponseEntity.status(HttpStatus.FORBIDDEN).build();

        URI location = ServletUriComponentsBuilder
                .fromCurrentRequest().path("/{id}")
                .buildAndExpand(message.getFromUser()).toUri();

        return ResponseEntity.created(location).build();
    }

    private List<String> getFriendsForUser(String username, String filter) {
        String url = "http://localhost:4567/friendships?username="
```

```
            + username + "&filter=" + filter;
        RestTemplate template = new RestTemplate();
        UserFriendships friendships = template.getForObject(url,
UserFriendships.class);
        return friendships.getFriendships();
    }
}
```

（2）新建一个名为 isFollowing 的方法，使用它替换掉原来的 getFriendsForUser 方法。我们给新方法一个@Async 注解，以告诉 Spring Boot，该方法将在另一个线程中运行。具体如下：

```
import org.springframework.scheduling.annotation.Async;
import java.util.concurrent.CompletableFuture;

...

@Async
public CompletableFuture<Boolean> isFollowing(String fromUser,
String toUser) {

    String url = String.format(
        "http://localhost:4567/followings?user=%s&filter=%s",
        fromUser, toUser);

    RestTemplate template = new RestTemplate();
    UserFollowings followings = template.forObject(url,
UserFollowings.class);

    return CompletableFuture.completedFuture(
        followings.getFollowings().isEmpty()
    );
}
```

（3）修改 create 方法以进行两个服务的调用。我们需要等到它们都完成后再决定是否继续，但是这两个服务的调用将同时进行。具体如下：

```
@RequestMapping(
        path="/messages",
        method=RequestMethod.POST,
        produces="application/json")
public ResponseEntity<Message> create(@RequestBody Message message) {
```

```
    CompletableFuture<Boolean> result1 =
isFollowing(message.getFromUser(), message.getToUser());
    CompletableFuture<Boolean> result2 =
isFollowing(message.getToUser(), message.getFromUser());
    CompletableFuture.allOf(result1, result2).join();

    // 如果二者均不为真,说明他们相互之间未加关注,因此返回 403
    if (!(result1.get() && result2.get()))
        ResponseEntity.status(HttpStatus.FORBIDDEN).build();

    ... // 继续
}
```

(4)由于 @Async 注解将在单独的线程上调度方法,因此需要配置一个 Executor。这可以在 Application 类中完成,具体如下:

```
package com.packtpub.microservices;

import org.springframework.boot.SpringApplication;
import org.springframework.boot.autoconfigure.SpringBootApplication;
import org.springframework.context.annotation.Bean;
import org.springframework.scheduling.annotation.EnableAsync;
import org.springframework.scheduling.concurrent.ThreadPoolTaskExecutor;

import java.util.concurrent.Executor;

@SpringBootApplication
@EnableAsync
public class Application {

    public static void main(String[] args) {
        SpringApplication.run(Application.class, args).close();
    }

    @Bean
    public Executor asyncExecutor() {
        ThreadPoolTaskExecutor executor = new ThreadPoolTaskExecutor();
        executor.setCorePoolSize(2);
        executor.setMaxPoolSize(2);
        executor.setQueueCapacity(500);
        executor.setThreadNamePrefix("SocialServiceCall-");
        executor.initialize();
```

```
            return executor;
    }
}
```

我们的服务现在会对 social-service 服务进行并发异步调用，以确保消息的发送者和接收者已经相互关注。我们自定义了包括 Executor 的 Async 调度程序，Executor 定义为应用程序配置的一部分。我们还配置了 ThreadPoolTaskExecutor 类，以将线程数限制为 2，队列大小限制为 500。在配置 Executor 时，要考虑许多因素，例如期望服务接收的流量总量，以及服务处理一个请求需要的平均时间等。在本示例中，我们保留了这些值。

3.4 使用服务发现来查找服务

在服务可以相互调用之前，它们需要使用某种服务发现（Service Discovery）机制，以便能够相互找到对方。这意味着需要将服务名称转换为网络位置（IP 地址和端口）。

3.4.1 理论阐释

传统应用程序找到服务的做法是，维护服务的网络位置，以将请求发送给它。这个网络位置可能是在配置文件中（或者采用了更糟的做法，将其硬编码到应用程序代码中——在 3.2 节"从服务到服务的通信"中就是这样做的）。这种方法假设网络位置是相对静态的，而在现代的云原生（Cloud-Native）应用程序中其实并非如此。微服务架构的拓扑在不断变化。节点是通过自动缩放来添加和删除的，我们必须假设某些节点很可能会完全失败，或者即使未失败，但是其处理请求的延迟也高到无法忍受。因此，随着微服务架构的发展，开发人员需要考虑一种功能更丰富的服务发现机制。

在选择服务发现机制时，用于备份服务注册表的数据存储非常重要。你需要一个经过充分测试且稳定可靠的系统。Apache ZooKeeper 是一个开源的分层键值存储，通常用于分布式锁定、服务发现、维护配置信息和其他分布式协调任务。ZooKeeper 的开发部分受 Google 于 2006 年发表的一篇论文的推动，该论文描述了 Chubby，这是一种内部开发的分布式锁存储系统。在本秘笈中，我们将使用 ZooKeeper 构建服务发现机制。

Spring Cloud ZooKeeper 是一个项目，可在 Spring Boot 应用程序中轻松实现 ZooKeeper 集成。

3.4.2 实战操作

本秘笈有两套步骤,所以分两部分进行介绍。

1. 在服务注册表中注册

在服务注册表中注册的秘笈需要一个正在运行的 ZooKeeper 集群。开发人员至少需要在开发计算机上以本地方式运行一个 ZooKeeper 节点。有关安装和运行 ZooKeeper 的详细指导,请阅读 ZooKeeper 的说明文档。本秘笈需要执行以下操作。

(1)在本示例中,我们将创建一个服务来处理用户账户的创建和检索。因此,可以使用以下 build.gradle 文件创建一个名为 users-service 的新 Gradle Java 应用程序:

```
group 'com.packtpub.microservices'
version '1.0-SNAPSHOT'

buildscript {
    repositories {
        mavenCentral()
    }
    dependencies {
        classpath group: 'io.spring.gradle', name: 'dependency-management-plugin', version: '0.5.6.RELEASE'
        classpath group: 'org.springframework.boot', name: 'spring-boot-gradle-plugin', version: '1.5.9.RELEASE'
    }
}

apply plugin: 'java'
apply plugin: 'org.springframework.boot'
apply plugin: "io.spring.dependency-management"

sourceCompatibility = 1.8

dependencyManagement {
    imports {
        mavenBom 'org.springframework.cloud:spring-cloud-zookeeper-dependencies:1.1.1.RELEASE'
    }
}

repositories {
```

```
    mavenCentral()
}

dependencies {
    compile group: 'io.reactivex', name: 'rxjava', version: '1.1.5'
    compile group: 'org.springframework.boot', name: 'spring-boot-starter-web'
    compile group: 'org.springframework.cloud', name: 'spring-cloud-starter-zookeeper-discovery', version: '1.1.1.RELEASE'
    testCompile group: 'junit', name: 'junit', version: '4.12'
}
```

（2）由于我们已经声明 spring-boot-starter-zookeeper-discovery 为依赖项，因此可以访问必要的注解，以告诉应用程序在启动时向 ZooKeeper 服务注册表进行注册。创建一个名为 Application 的新类，它将作为我们服务的入口点，代码如下：

```
package com.packtpub.microservices.ch03.servicediscovery;

import org.springframework.boot.SpringApplication;
import org.springframework.boot.autoconfigure.SpringBootApplication;
import org.springframework.cloud.client.discovery.EnableDiscoveryClient;

@EnableDiscoveryClient
@SpringBootApplication
public class Application {
    public static void main(String[] args) {
        SpringApplication.run(Application.class, args);
    }
}
```

（3）现在，应用程序将会尝试连接到 ZooKeeper 节点，默认情况下，该节点运行在 localhost 的端口 2181 上。请注意，此默认设置仅适用于本地开发，当部署到生产环境中时，必须进行更改。添加包含以下内容的文件 src/resources/application.yml：

```
spring:
    cloud:
        zookeeper:
            connect-string: localhost:2181
```

（4）要在服务注册表中为服务提供有意义的名称，请修改 application.yml 文件并添加以下内容：

```
spring:
    cloud:
        zookeeper:
            connect-string: localhost:2181
    application:
        name: users-service
```

2. 寻找服务

现在，我们已经在服务注册表中注册了一个服务，接下来将创建另一个服务来演示如何使用 Spring ZooKeeper DiscoveryClient 查找该服务的运行实例。

（1）打开先前创建的消息服务客户端。将以下行添加到 build.gradle 中：

```
group 'com.packtpub.microservices'
version '1.0-SNAPSHOT'

buildscript {
    repositories {
        mavenCentral()
    }
    dependencies {
        classpath group: 'io.spring.gradle', name: 'dependency-management-plugin', version: '0.5.6.RELEASE'
        classpath group: 'org.springframework.boot', name: 'spring-boot-gradle-plugin', version: '1.5.9.RELEASE'
    }
}

apply plugin: 'java'
apply plugin: 'org.springframework.boot'
apply plugin: 'io.spring.dependency-management'

sourceCompatibility = 1.8

dependencyManagement {
    imports {
        mavenBom 'org.springframework.cloud:spring-cloud-zookeeper-dependencies:1.1.1.RELEASE'
    }
}

repositories {
    mavenCentral()
```

```
}
dependencies {
    compile 'io.reactivex:rxjava:1.3.4'
    compile group: 'org.springframework.cloud', name: 'spring-
cloud-starter-zookeeper-discovery', version: '1.1.1.RELEASE'
    compile group: 'org.springframework.cloud', name: 'spring-
cloud-starter-feign', version: '1.2.5.RELEASE'
    compile group: 'org.springframework.kafka', name: 'spring-
kafka', version: '2.1.1.RELEASE'
    compile group: 'org.springframework.boot', name: 'spring-boot-
starter-web'
    testCompile group: 'junit', name: 'junit', version: '4.12'
}
```

(2)我们使用的是 Netflix 开发的 HTTP 客户端,称为 Feign。Feign 允许你声明性地构建 HTTP 客户端,并且默认情况下支持服务发现。据此,创建一个名为 UsersClient.java 的新文件,其内容如下:

```
package com.packtpub.microservices.ch03.servicediscovery.clients;

import org.springframework.beans.factory.annotation.Autowired;
import org.springframework.cloud.client.discovery.EnableDiscoveryClient;
import org.springframework.cloud.netflix.feign.EnableFeignClients;
import org.springframework.cloud.netflix.feign.FeignClient;
import org.springframework.context.annotation.Configuration;
import org.springframework.web.bind.annotation.PathVariable;
import org.springframework.web.bind.annotation.RequestMapping;
import org.springframework.web.bind.annotation.RequestMethod;
import org.springframework.web.bind.annotation.ResponseBody;

import java.util.List;

@Configuration
@EnableFeignClients
@EnableDiscoveryClient
public class UsersClient {

    @Autowired
    private Client client;

    @FeignClient("users-service")
    interface Client {
```

```
    @RequestMapping(path = "/followings/{userId}", method =
RequestMethod.GET)
    @ResponseBody
    List<String> getFollowings(@PathVariable("userId") String
userId);
    }

    public List<String> getFollowings(String userId) {
        return client.getFollowings(userId);
    }
}
```

(3)打开 MessageController.java 文件,并向该文件中添加一个 UsersClient 实例作为字段,具体如下:

```
package com.packtpub.microservices;
...
@RestController
public class MessagesController {
    ...
    @Autowired
    private UsersClient usersClient;
    ...
}
```

(4)现在,我们不必在 isFollowing 方法中手动构建 URL,而是可以使用 Feign 客户端自动获取用户的好友关系列表,具体如下:

```
@Async
public CompletableFuture<Boolean> isFollowing(String fromUser,
String toUser) {
    List<String> friends = usersClient.getFollowings(fromUser)
            .stream()
            .filter(toUser::equals)
            .collect(Collectors.toList());

    return CompletableFuture.completedFuture(friends.isEmpty());
}
```

因为我们现在使用的是服务注册表,所以再也不必担心以前那样笨拙的配置出现的问题(配置文件保留的主机名的值很可能会发生变化导致服务不可用)。此外,我们还可以开始决定如何在服务的可用实例之间分配负载。

3.5 服务器端负载均衡

当考虑在运行应用程序实例的服务器集群之间分配负载时，不妨先来回溯 Web 应用程序架构的发展历史。最初，很多 Web 应用程序都是由 Web 服务器（如 Apache 或类似的 Web 服务器守护程序软件）托管的静态 HTML 页面；后来，开发人员使用了诸如通过 CGI 执行的服务器端脚本之类的技术，应用程序才逐渐变得更加动态，但是动态应用程序仍然是由 Web 服务器守护程序直接托管和提供服务的文件。这种简单的架构维持了很长一段时间。但是，最终随着应用程序接收到的流量不断增加，开发人员需要找到一种在应用程序的相同无状态实例之间分配负载的方法。

3.5.1 理论阐释

在负载均衡方面，当前也有许多技术，如轮询（Round-Robin）DNS 或 DNS 地理位置。微服务负载均衡最简单、最常见的形式是使用软件程序将请求转发到后端服务器集群的服务器之一。根据我们选择的负载均衡器使用的特定负载均衡算法，可以采用多种不同的方式分配负载。简单的负载均衡算法包括轮询和随机选择两种方式。在现实世界的生产应用程序中，我们通常会选择一种负载均衡算法，该算法在选择集群中的节点以便将请求转发给它时，将会考虑已经报告的度量（如负载或活动连接数）。

有许多流行的开源应用程序可以对微服务执行有效的负载均衡。例如，HAProxy 就是一种流行的开源负载均衡器，它可以执行 TCP 和 HTTP 负载均衡。NGINX 是一种流行的开源 Web 服务器，可以有效地用作反向代理、应用程序服务器、负载均衡器，甚至 HTTP 缓存。如今，越来越多的组织都需要开发部署在云平台（如 Amazon Web Services 或 Google Cloud Platform）上的微服务，而这些云平台各自具有用于服务器端负载均衡的解决方案。

AWS 提供了一种称为弹性负载均衡（Elastic Load Balancing，ELB）的负载均衡解决方案。ELB 可以被配置为将流量转发到自动缩放组（Auto Scale Group，ASG）的成员。自动缩放组（ASG）是被视为逻辑组的 EC2 实例的集合。ELB 使用运行状况检查（TCP 或 HTTP）来帮助负载均衡器确定是否将流量转发到特定的 EC2 实例。

在本秘笈中，我们将使用 AWS CLI 工具创建自动缩放组（ASG）并将 ELB 附加到该组中。我们不会在本秘笈中介绍配置管理或部署，因此，假设你已经在自动缩放组中的每个 EC2 实例上都运行了一个微服务。

3.5.2 实战操作

本秘笈需要执行以下操作。

（1）在本秘笈中，我们将使用 AWS CLI，这是一个用 Python 编写的命令行实用程序，它可以简化与 AWS API 的交互。我们还假设你已经拥有一个 AWS 账户，并且已经安装和配置了 AWS CLI 应用程序。有关 AWS CLI 安装和配置的详细指导，可访问以下网址中的 AWS 说明文档：

https://docs.aws.amazon.com/cli/latest/index.html#

（2）创建启动配置。启动配置是自动缩放组将用于创建新 EC2 实例的模板。它们包含诸如在创建新实例时要使用的实例类型和大小之类的信息。开发人员可以为启动配置指定一个唯一的名称。在本示例中，直接就将其称为 users-service-launch-configuration：

```
$ aws create-launch-configuration --launch-configuration-name users-service-launch-configuration \
    --image-id ami-05355a6c --security-groups sg-8422d1eb \
    --instance-type m3.medium
```

（3）创建一个使用新启动配置的自动缩放组，具体如下：

```
$ aws create-auto-scaling-group --auto-scaling-group-name users-service-asg \
    --launch-configuration-name users-service-launch-configuration \
    --min-size 2 \
    --max-size 10
```

（4）创建一个 ELB，具体如下：

```
$ aws create-load-balancer --load-balancer-name users-service-elb \
    --listeners "Protocol=HTTP,LoadBalancerPort=80,InstanceProtocol=HTTP,InstancePort=8080"
```

（5）通过运行以下命令行将 ASG 附加到负载均衡器中：

```
$ aws autoscaling attach-load-balancers --auto-scaling-group-name users-service-asg --load-balancer-names users-service-elb
```

3.6 客户端负载均衡

服务器端负载均衡是将负载分配到应用程序的一种行之有效的方法。但是，它的缺点在于，单台负载均衡器可以处理的传入连接数有上限。这至少可以通过轮询 DNS 部分解决，轮询 DNS 会将负载分配给多台负载均衡器，但是这样的配置很快就会变得既烦琐又昂贵。在本已非常复杂的微服务架构中，负载均衡器应用程序也可能成为故障点。

3.6.1 理论阐释

有鉴于服务器端负载均衡存在的一些缺陷，客户端负载均衡正逐渐成为取代服务器端负载均衡的一种越来越流行的备选方法。在此约定中，客户端负责将请求平均分配给正在运行的服务实例。客户端可以跟踪来自节点的延迟和故障率，并选择减少流向遭遇到高延迟或高故障率的节点的流量。这种负载均衡方法非常有效和简单，特别是在大规模应用中。

Ribbon 是由 Netflix 开发的开源库，除其他功能外，它还提供对客户端负载均衡的支持。在本秘笈中，我们将修改 message-service 服务，以便将 Ribbon 用于客户端负载均衡。我们不会将对用户好友关系的请求发送给单个用户服务实例，而是将负载分配给多个可用实例。

3.6.2 实战操作

本秘笈需要执行以下操作。

（1）打开 message-service 项目，并将以下代码添加到 build.gradle 中：

```
...
dependencies {
    ...
    compile group: 'org.springframework.cloud', name: 'spring-cloud-starter-ribbon', version: '1.4.2.RELEASE'
}
...
```

（2）导航到 src/main/resources/application.yml 并为 users-service 添加以下配置：

```
users-service:
    ribbon:
```

```yaml
    eureka:
        enabled: false
    listOfServers: localhost:8090,localhost:9092,localhost:9999
    ServerListRefreshInterval: 15000
```

（3）创建一个名为 UsersServiceConfiguration 的新 Java 类。该类将配置 ribbon 在决定如何分配负载时应遵循的特定规则：

```java
package com.packtpub.microservices.ch03.clientsideloadbalancing;

import org.springframework.beans.factory.annotation.Autowired;
import org.springframework.context.annotation.Bean;

import com.netflix.client.config.IClientConfig;
import com.netflix.loadbalancer.IPing;
import com.netflix.loadbalancer.IRule;
import com.netflix.loadbalancer.PingUrl;
import com.netflix.loadbalancer.AvailabilityFilteringRule;

public class UsersServiceConfiguration {

    @Autowired
    IClientConfig ribbonClientConfig;

    @Bean
    public IPing ribbonPing(IClientConfig config) {
        return new PingUrl();
    }

    @Bean
    public IRule ribbonRule(IClientConfig config) {
        return new AvailabilityFilteringRule();
    }

}
```

（4）打开 MessageController 并将以下注解添加到 MessageController 类中：

```java
@RibbonClient(name = "users-service", configuration = UsersServiceConfiguration.class)
@RestClient
public class MessageController {
```

（5）注解 RestTemplate 类，以表明我们希望它使用 ribbon 负载均衡支持，并修改 URL 以使用服务名称，而不是我们之前硬编码的主机名。具体如下：

```
@RibbonClient(name = "users-service", configuration = UsersServiceConfiguration.class)
@RestClient
public class MessageController {
    ...
    @LoadBalanced
    @Bean
    RestTemplate restTemplate(){
        return new RestTemplate();
    }
    ...

    @Async
    public CompletableFuture<Boolean> isFollowing(String fromUser, String toUser) {

        String url = String.format(
"http://localhost:4567/followings?user=%s&filter=%s",
            fromUser, toUser);

        RestTemplate template = new RestTemplate();
        UserFriendships followings = template.getForObject(url, UserFriendships.class);

        return CompletableFuture.completedFuture(
            followings.getFriendships().isEmpty()
        );
    }
}
```

3.7 构建事件驱动的微服务

到目前为止，我们所有的服务到服务通信秘笈都涉及一个服务直接调用一个或多个其他服务。当需要来自下游服务的响应来满足用户的请求时，这是必需的。但是，这并不总是必需的。如果你想对系统中的事件做出反应，例如，当你要发送电子邮件或通知，或者要更新分析存储时，最好使用事件驱动的架构。

3.7.1 理论阐释

在事件驱动设计中，一个服务将生成一条消息给代理，而另一个应用将使用该消息并执行操作。这样做的好处是可以将消息的发布者（Publisher）与消息的使用者（Consumer）分离（例如，你的 message-service 服务就不必担心发送电子邮件通知的事情），并且还可以消除用户请求的关键路径上的潜在昂贵操作。像这样的事件驱动架构还提供了一定程度的容错能力，因为消息的使用者可能会操作失败，在这种情况下就可以重放消息以重试任何失败的操作。

Apache Kafka 是一个开源流处理平台。它的核心是一个事件代理，其构造为分布式事务日志。如果要详细介绍 Apache Kafka，那么完全可以写出一本厚厚的图书，因此，如果开发人员想要获得有关它的更多信息，我们强烈建议你阅读 LinkedIn（领英）上介绍 Kafka 的博客文章，其网址如下：

https://engineering.linkedin.com/distributed-systems/log-what-every-software-engineer-should-know-about-real-time-datas-unifying

要理解和掌握本秘笈的操作，你至少需要知道，Kafka 是一个分布式事件存储，它允许你将消息发布到称为主题（Topic）的类别中，然后，另一个进程可以使用主题中的消息并对消息做出反应。

回到我们虚构的图像消息应用程序中，当一个用户向另一个用户发送消息时，我们希望能够以多种方式通知接收者。根据接收者的喜好配置，我们可能会发送电子邮件或推送通知，或同时发送这二者。在本秘笈中，我们将修改以前秘笈中的 message-service 服务，以将事件发布到称为 message 的 Kafka 主题中。然后，我们将构建一个使用者应用程序，该应用程序将侦听消息主题中的事件，并可以通过发送接收者通知来做出响应。

3.7.2 实战操作

用于 Apache Kafka 的 Spring（spring-kafka）是一个可以轻松将 Spring 应用程序与 Apache Kafka 集成的项目。它为发送和接收消息提供了有用的抽象。

请注意，要按照本秘笈中的步骤操作，你将需要运行并可以访问某个版本的 Kafka 和 ZooKeeper。安装和配置这两个软件不在本秘笈的讨论范围之内，因此请访问相应的项目网站，并按照其详细的入门指南进行操作。在本秘笈中，我们假设 Kafka 在端口 9092 上运行单个代理，同时在端口 2181 上运行 ZooKeeper 的单个实例。

1. 消息生产者

(1) 打开之前秘笈中编写的 message-service 项目。修改 build.gradle 文件,并将 spring-kafka 项目添加到依赖项列表中,具体如下:

```
dependencies {
    compile group: 'org.springframework.kafka', name: 'spring-kafka', version: '2.1.1.RELEASE'
    compile group: 'org.springframework.boot', name: 'spring-boot-starter-web'
    testCompile group: 'junit', name: 'junit', version: '4.12'
}
```

(2) spring-kafka 项目提供了用于向 Kafka 代理发送消息的模板。要在项目中使用模板,需要创建一个 ProducerFactory 接口并将其提供给模板的构造函数。

(3) 打开 Application.java 文件并添加以下内容。请注意,我们在这里对 Kafka 代理的网络位置进行了硬编码,但是在实际应用中,你至少应将此值放在某种配置中(最好遵循十二因素的约定):

```
package com.packtpub.microservices.ch03.message;

import org.apache.kafka.clients.producer.ProducerConfig;
import org.apache.kafka.common.serialization.StringSerializer;
import org.springframework.boot.SpringApplication;
import org.springframework.boot.autoconfigure.SpringBootApplication;
import org.springframework.context.annotation.Bean;
import org.springframework.kafka.core.DefaultKafkaProducerFactory;
import org.springframework.kafka.core.KafkaTemplate;
import org.springframework.kafka.core.ProducerFactory;

import java.util.HashMap;
import java.util.Map;

@SpringBootApplication
@EnableAsync
public class Application {
    public static void main(String[] args) {
        SpringApplication.run(Application.class, args);
    }

    @Bean
```

```
    public Map<String, Object> producerConfigs() {
        Map<String, Object> props = new HashMap<>();
        props.put(ProducerConfig.BOOTSTRAP_SERVERS_CONFIG,
"localhost:9092");
        props.put(ProducerConfig.KEY_SERIALIZER_CLASS_CONFIG,
StringSerializer.class);
        props.put(ProducerConfig.VALUE_SERIALIZER_CLASS_CONFIG,
StringSerializer.class);
        return props;
    }

    @Bean
    public ProducerFactory<Integer, String> producerFactory() {
        return new
DefaultKafkaProducerFactory<>(producerConfigs());
    }

    @Bean
    public KafkaTemplate<Integer, String> kafkaTemplate() {
        return new KafkaTemplate<Integer,
String>(producerFactory());
    }
}
```

（4）现在可以在应用程序中使用 KafkaTemplate，我们将其中一个模板添加到 MessageController 类中。另外，还可以使用 Jackson ObjectMapper 类将我们的 Message 实例转换为将发布到 Kafka 主题的 JSON 字符串。

打开 MessageController 类，并添加以下字段：

```
...
import org.springframework.kafka.core.KafkaTemplate;
import com.fasterxml.jackson.databind.ObjectMapper;
...


@RestController
public class MessageController {

    @Autowired
    private KafkaTemplate kafkaTemplate;

    @Autowired
```

```
    private ObjectMapper objectMapper;

    ...
}
```

（5）现在可以访问 Jackson ObjectMapper 和 KafkaTemplate 类，创建一个用于发布事件的方法。在此示例中，我们将输出到 System.err（标准错误）和 System.out（标准输出）中。在实际的应用程序中，你应该配置一个日志管理程序，如 log4j，并使用适当的日志级别，具体如下：

```
@RestController
public class MessageController {

    ...

    private void publishMessageEvent(Message message) {
        try {
            String data = objectMapper.writeValueAsString(message);
            ListenableFuture<SendResult> result = kafkaTemplate.send("messages", data);
            result.addCallback(new ListenableFutureCallback<SendResult>() {
                @Override
                public void onFailure(Throwable ex) {
                    System.err.println("Failed to emit message event: " + ex.getMessage());
                }

                @Override
                public void onSuccess(SendResult result) {
                    System.out.println("Successfully published message event");
                }
            });
        } catch (JsonProcessingException e) {
            System.err.println("Error processing json: " + e.getMessage());
        }
    }
}
```

（6）将以下行添加到 create 方法中，调用先前创建的 publishMessageEvent 方法：

```
@RequestMapping(
        path="/messages",
        method=RequestMethod.POST,
        produces="application/json")
public ResponseEntity<Message> create(@RequestBody Message message)
{
    ...
    publishMessageEvent(message);
    return ResponseEntity.created(location).build();
}
```

(7)要测试此示例,请使用 kafka-topics.sh 这个 Kafka 实用程序(与 Kafka 二进制发行版打包在一起)创建消息主题,具体如下:

```
bin/kafka-topics.sh --create \
    --zookeeper localhost:2181 \
    --replication-factor 1 --partitions 1 \
    --topic messages
```

2. 消息使用者

在发布了消息发送事件之后,下一步就是构建一个小型使用者应用程序,它可以对系统中的这些事件做出反应。在本秘笈中,我们将仅讨论与 Kafka 有关的操作,至于电子邮件和推送通知功能的实现,则留给读者作为练习。

(1)创建一个名为 message-notifier 的新 Gradle Java 项目,包括以下 build.gradle 文件:

```
group 'com.packtpub.microservices'
version '1.0-SNAPSHOT'

buildscript {
    repositories {
        mavenCentral()
    }
    dependencies {
        classpath group: 'org.springframework.boot', name: 'spring-boot-gradle-plugin', version: '1.5.9.RELEASE'
    }
}

apply plugin: 'java'
apply plugin: 'org.springframework.boot'
```

第 3 章 服务间通信

```
sourceCompatibility = 1.8

repositories {
    mavenCentral()
}

dependencies {
    compile group: 'org.springframework.kafka', name: 'spring-kafka', version: '2.1.1.RELEASE'
    compile group: 'org.springframework.boot', name: 'spring-boot-starter'
    testCompile group: 'junit', name: 'junit', version: '4.12'
}
```

(2)使用 Spring Boot 应用程序样板创建一个名为 Application 的新 Java 类，具体如下：

```
package com.packtpub.microservices.ch03.consumer;

import org.springframework.boot.SpringApplication;
import org.springframework.boot.autoconfigure.SpringBootApplication;

@SpringBootApplication
public class Application {
    public static void main(String[] args) {
        SpringApplication.run(Application.class, args);
    }
}
```

3.8 不断演变的 API

API 是客户端和服务器之间的合同。对 API 进行向后不兼容的更改可能导致服务的客户端出现意外错误。在微服务架构中，必须采取预防措施以确保对服务 API 的更改不会在整个系统中无意间导致一连串的问题。

3.8.1 理论阐释

为了解决这种兼容性问题，一种流行的方法是对 API 进行版本控制，这既可以通过 URL 执行，也可以通过请求标头中的内容协商执行。因为 URL 前缀或查询字符串通常更易于使用，并且易于缓存，所以它们更为常见。在这种情况下，API 端点要么以版本字符串作为前缀（如/v1/users），要么使用查询字符串参数指定版本或日期以进行调用（如/v1/users?version=1.0 或/v1/users?version=20210122）。

使用边缘代理或服务网格配置，甚至可以在环境中运行多个版本的软件，并根据 URL 将请求路由到服务的旧版本或新版本。这改变了服务的传统生命周期——当某个版本不再接收到任何流量时，你可以安全地停用该版本。这种策略可能很有用，尤其是在公共 API 中，你几乎无法控制客户端的情况下。

微服务不同于公共 API。客户端和服务器之间使用公共 API 的合同寿命更长，而在微服务架构中，更容易追踪正在使用你服务的客户端并说服他们升级其代码。

尽管如此，有时还是需要 API 版本控制。由于能够成功响应多个 API 版本是一项维护上的负担，因此，我们仍希望尽可能地避免使用它。为此，可以使用一些策略来避免进行向后不兼容的更改。

3.8.2 实战操作

本秘笈需要执行以下操作。

（1）使用示例应用程序 pichat，假设我们想要将消息正文的名称从 body 更改为 message_text。这是一个问题，因为我们的 message-service 服务被设计为接收以下请求：

```
GET /messages?user_id=123
GET /messages/123
POST /messages
DELETE /messages/123
```

（2）对于 GET 请求来说，客户端将在响应中包含一个带有称为 body 的字段的 JSON 对象；对于 POST 请求来说，客户端将使用称为 body 的字段将有效负载作为 JSON 对象发送。因此，不能简单地删除 body（这会破坏现有的客户端），而只能更改 API 版本。我们将在旧字段之外添加新字段，具体如下：

```
{
    "message": {
        "from_user": "sender",
        "to_user": "recipient",
        "body": "Hello, there",
        "message_text": "Hello, there"
    }
}
```

（3）现在，开发人员可以通过这些响应字段逐步跟踪客户端。当客户端全部升级后，即可安全地从 JSON 响应中删除那些已经被弃用的字段。

第 4 章　客户端模式

本章包含以下操作秘笈。
- ❏　使用依赖性的 Future 对并发进行建模。
- ❏　服务于前端的后端。
- ❏　使用 JSON 和 HTTP 实现 RPC 一致性。
- ❏　使用 Thrift。
- ❏　使用 gRPC。

4.1　导　　语

在构建面向服务的架构时，很容易陷入一种思考泥潭，即如何采用最通用的方式来表示由特定服务控制的域实体和行为。但事实是，我们很少以通用方式使用服务——我们通常将对多个服务的调用组合在一起，并使用响应来创建新的聚合响应主体。

我们通常以类似于过去从数据库聚合数据的方式进行服务调用，因此必须考虑系统中不同类型之间的关系以及如何以最佳方式对数据依赖关系进行建模。

我们还希望使客户端的开发变得更容易。在设计通用 API 时，也有一个容易陷入的思考泥潭，那就是如何以正确的方法做事而不是以简单的方式做事。例如，你也许听到过有人批评某个 API 设计不是 RESTful 架构的，这种论调就很可能已经落入这种窠臼。对于某个服务来说，如果客户端需要对它进行数十次的调用才能获取所需的数据，那么该服务就难言优秀。在设计包含微服务的系统时，必须从客户端的角度考虑数据聚合。

客户端不仅要考虑其正在调用的服务，而且往往还必须考虑到这些服务的实例，它们需要配置自身才能进行调用。常见的是使用模拟环境（Staging Environment）或测试环境（Testing Environment）。在微服务架构中，这些环境变得更加复杂。

在本章中，我们将讨论用于对依赖性的服务调用进行建模的技巧，以及如何聚合来自各种服务的响应以创建特定于客户端的 API。我们还将讨论如何管理不同的微服务环境、如何使 RPC 与 JSON 和 HTTP 保持一致，以及 gRPC 和 Thrift 二进制协议的应用。

4.2　使用依赖性的 Future 对并发进行建模

在前面的秘笈中已经介绍过，开发人员可以使用异步方法进行服务调用，这些服务调用在单独的线程中处理。这非常重要,因为阻塞网络的输入/输出(I/O)将严重限制我们的服务能够处理的传入请求的数量。

4.2.1 理论阐释

如果服务会阻塞网络 I/O,那么它在每个进程中就只能处理相对较少的请求,这要求我们在水平扩展上花费更多的资源。在我们的图像消息示例中,message-service 服务需要为两个用户(消息的发送者和接收者)调用 social-service 社交图服务,确保在允许发送消息之前,这两个用户彼此已添加关注(或称为"互粉")。我们修改了请求方法,以返回包装响应的 CompletableFuture 实例,然后等待所有结果完成,以验证消息的发送者和接收者是否具有对称的互粉关系。当你发出多个互不依赖的请求(也就是说,你不需要获得上一个请求的响应即可发出下一个请求)时,此模块应该能够正常工作。但是,在这种情况下,如果我们有依赖性的服务调用,则需要一种更好的方法来对依赖进行建模。

例如,在我们假想的 pichat 应用程序中,我们需要显示一个屏幕,其中列出了已经关注的用户的信息。为此,我们需要调用 social-service 的社交图服务以获取用户列表,然后调用 users-service 服务以获取用户的详细信息,如每个用户的显示名称和头像。这样的用例就涉及进行有依赖性的服务调用。我们需要一种有效的方法来对这种服务调用进行建模,同时仍然以异步操作方式来调度,允许它们在单独的执行线程中运行。

在本秘笈中,我们将通过使用 CompletableFuture 以及 Java 8 Stream(流)的组合来对依赖性的服务调用进行建模,进而演示上述操作。我们将创建一个示例客户端应用程序,该应用程序将调用 social-service 服务以获取已登录用户所关注的用户列表,然后调用 users-service 服务以获取每个用户的详细信息。

4.2.2 实战操作

为了对依赖性的异步服务调用进行建模,我们将需要利用 Java 8 中的两个功能。流对于处理数据很有用,因此在本示例中,将使用它们来从已经加关注的列表中提取用户名,并将函数映射到每个元素。Java 8 的 CompletableFuture 是可以组合的,这使我们可以自然地表达 Future 之间的依赖关系。

在此秘笈中,我们将创建一个简单的客户端应用程序,该应用程序将调用 social-service 服务以获取当前用户已经加关注的用户列表。对于返回的每个用户,应用程序将从 users-service 服务中获取用户的详细信息。为了便于演示,我们将本示例构建为命令行应用程序,但它也可以是另一个微服务、网页或移动客户端。

> **注意：**
> 为了构建具有 Spring Boot 应用程序所有功能的命令行应用程序，我们将采取一些作弊手段，仅实现 CommandLineRunner 并在 run()方法中调用 System.exit(0);。

在开始构建应用程序之前，我们将简要介绍假设的 social-service 服务和 users-service 服务的响应。此外，我们还可以通过仅将适当的 JSON 响应托管在本地 Web 服务器上来模拟这些服务。最后我们还将分别使用端口 8000 和 8001 来运行 social-service 服务和 users-service 服务。social-service 服务有一个端点/followings/:username，它将返回一个 JSON 对象，其中包含指定用户名的已经加关注的好友列表。JSON 响应将类似于以下片段：

```
{
    "username": "paulosman",
    "followings": [
        "johnsmith",
        "janesmith",
        "petersmith"
    ]
}
```

users-service 服务具有一个名为/users/:username 的端点，它将返回用户详细信息的 JSON 表示，包括用户名、全名和头像 URL 等，具体如下：

```
{
    "username": "paulosman",
    "full_name": "Paul Osman",
    "avatar_url": "http://foo.com/pic.jpg"
}
```

现在我们已经有了自己的服务，并且简要介绍了希望从每个服务获得的响应，接下来即可继续执行以下步骤来构建客户端应用程序。

（1）创建一个名为 UserDetailsClient 的新 Java/Gradle 应用程序。其 build.gradle 文件将包含以下内容：

```
group 'com.packtpub.microservices'
version '1.0-SNAPSHOT'

buildscript {
    repositories {
        mavenCentral()
    }
    dependencies {
```

```
        classpath group: 'org.springframework.boot', name: 'spring-boot-gradle
        -plugin', version: '1.5.9.RELEASE'
    }
}

apply plugin: 'java'
apply plugin: 'org.springframework.boot'

sourceCompatibility = 1.8

repositories {
    mavenCentral()
}

dependencies {
    testCompile group: 'junit', name: 'junit', version: '4.12'
    compile group: 'org.springframework.boot',
    name: 'spring-boot-starter-web'
}
```

（2）创建一个名为 com.packtpub.microservices.ch04.user.models 的程序包和一个名为 UserDetails 的新类。我们将使用该类来对用户服务的响应进行建模，具体如下：

```
package com.packtpub.microservices.ch04.user.models;

import com.fasterxml.jackson.annotation.JsonProperty;

public class UserDetails {
    private String username;

    @JsonProperty("display_name")

    private String displayName;

    @JsonProperty("avatar_url")
    private String avatarUrl;

    public UserDetails() {}

    public UserDetails(String username, String displayName,
    String avatarUrl) {
        this.username = username;
```

```
        this.displayName = displayName;
        this.avatarUrl = avatarUrl;
    }

    public String getUsername() {
        return username;
    }

    public void setUsername(String username) {
        this.username = username;
    }

    public String getDisplayName() {
        return displayName;
    }

    public void setDisplayName(String displayName) {
        this.displayName = displayName;
    }

    public String getAvatarUrl() {
        return avatarUrl;
    }

    public void setAvatarUrl(String avatarUrl) {
        this.avatarUrl = avatarUrl;
    }

    public String toString() {
        return String.format("[UserDetails: %s, %s, %s]", username,
        displayName, avatarUrl);
    }
}
```

（3）在 com.packtpub.microservices.ch04.user.models 程序包中创建另一个类，名为 Followings。这将用于对 social-service 服务的响应进行建模，具体如下：

```
package com.packtpub.microservices.ch04.user.models;

import java.util.List;

public class Followings {
    private String username;
```

```java
    private List<String> followings;

    public Followings() {}

    public Followings(String username, List<String> followings) {
        this.username = username;
        this.followings = followings;
    }

    public String getUsername() {
        return username;
    }

    public void setUsername(String username) {
        this.username = username;
    }

    public List<String> getFollowings() {
        return followings;
    }

    public void setFollowings(List<String> followings) {
        this.followings = followings;
    }

    public String toString() {
        return String.format("[Followings for username: %s - %s]",
            username, followings);
    }
}
```

（4）创建一个服务表示来调用 social-service。为了更好地体现其意义，我们将其命名为 SocialService，并将它放在 com.packtpub.microservices.ch04.user.services 程序包中，具体如下：

```java
package com.packtpub.microservices.ch04.user.services;

import com.packtpub.microservices.models.Followings;
import org.springframework.boot.web.client.RestTemplateBuilder;
import org.springframework.scheduling.annotation.Async;
import org.springframework.stereotype.Service;
import org.springframework.web.client.RestTemplate;
```

```java
import java.util.concurrent.CompletableFuture;

@Service
public class SocialService {

    private final RestTemplate restTemplate;

    public SocialService(RestTemplateBuilder restTemplateBuilder) {
        this.restTemplate = restTemplateBuilder.build();
    }

    @Async
    public CompletableFuture<Followings>
    getFollowings(String username) {
        String url =
String.format("http://localhost:8000/followings/
        %s", username);
        Followings followings = restTemplate.getForObject(url,
        Followings.class);
        return CompletableFuture.completedFuture(followings);
    }
}
```

（5）为 users-service 服务创建服务表示。同样，为了更好地体现其意义，我们将在同一个程序包中调用类 UserService，具体如下：

```java
package com.packtpub.microservices.services;

import com.packtpub.microservices.models.Followings;
import com.packtpub.microservices.models.UserDetails;
import org.springframework.boot.web.client.RestTemplateBuilder;
import org.springframework.scheduling.annotation.Async;
import org.springframework.stereotype.Service;
import org.springframework.web.client.RestTemplate;

import java.util.concurrent.CompletableFuture;

@Service
public class UserService {
    private final RestTemplate restTemplate;

    public UserService(RestTemplateBuilder restTemplateBuilder) {
        this.restTemplate = restTemplateBuilder.build();
```

```
    }

    @Async
    public CompletableFuture<UserDetails>
    getUserDetails(String username) {
        String url = String.format("http://localhost:8001/users/
        %s", username);
        UserDetails userDetails = restTemplate.getForObject(url,
        UserDetails.class);
        return CompletableFuture.completedFuture(userDetails);
    }
}
```

（6）现在，我们已经有了对服务的响应进行建模的类，并且也有了服务对象来表示我们将要调用的服务。现在可以通过创建主类将这些联系在一起，该主类将使用 Future 的可组合性对依赖关系进行建模，从而以依赖性的方式调用这两个服务。

创建一个名为 UserDetailsClient 的新类，具体如下：

```
package com.packtpub.microservices.ch04.user;

import com.packtpub.microservices.models.Followings;
import com.packtpub.microservices.models.UserDetails;
import com.packtpub.microservices.services.SocialService;
import com.packtpub.microservices.services.UserService;
import org.springframework.beans.factory.annotation.Autowired;
import org.springframework.boot.CommandLineRunner;
import org.springframework.boot.SpringApplication;
import org.springframework.boot.autoconfigure.SpringBootApplication;

import java.util.List;
import java.util.concurrent.CompletableFuture;
import java.util.concurrent.Future;
import java.util.stream.Collectors;

@SpringBootApplication
public class UserDetailsClient implements CommandLineRunner {

    public UserDetailsClient() {}

    @Autowired
    private SocialService socialService;

    @Autowired
```

```
    private UserService userService;

    public CompletableFuture<List<UserDetails>>
    getFollowingDetails(String username) {
        return socialService.getFollowings(username).thenApply(f ->
                f.getFollowings().stream().map(u ->userService.
                getUserDetails(u)).map(CompletableFuture::join).
                collect(Collectors.toList()));
    }

    public static void main(String[] args) {
        SpringApplication.run(UserDetailsClient.class, args);
    }

    @Override
    public void run(String... args) throws Exception {
        Future<List<UserDetails>> users = getFollowingDetails
        ("paulosman");
        System.out.println(users.get());
        System.out.println("Heyo");
        System.exit(0);
    }
}
```

该操作的秘密就在以下方法中：

```
CompletableFuture<List<UserDetails>> getFollowingDetails(String
username)
{
    return socialService.getFollowings(username).thenApply(
        f -> f.getFollowings().stream().map(u ->
            userService.getUserDetails(u)).map(
                CompletableFuture::join).collect(Collectors.toList()));
}
```

回想一下，SocialService 中的 getFollowings 方法返回 CompletableFuture<Followings>。CompletableFuture 具有一个名为 thenApply 的方法，该方法将获取 Future 的最终结果（Followings）并将其应用到 Lambda 中。在这种情况下，我们将采用 Followings，并使用 Java 8 Stream API 调用由 social-service 服务返回的用户名列表的映射。该映射将对每个用户名应用函数，函数将在 UserService 上调用 getUserDetails。CompletableFuture:: join 方法用于将 List<Future<T>>转换为 Future<List<T>>，这是执行此类依赖服务调用时的常

见操作。最后，我们收集结果并将其作为列表返回。

4.3 服务于前端的后端

当软件从桌面和基于 Web 的应用程序转移到移动 App 时，分布式架构变得更加流行。许多组织将重点放在构建平台而不是产品上。这种方法强调 API 的重要性，产品可以向客户端和第三方合作伙伴公开这些 API。

4.3.1 理论阐释

随着 API 成为任何基于 Web 的应用程序的给定条件，尝试基于相同的 API 构建客户端应用程序（移动应用或 JavaScript）的做法也变得很流行，这些 API 可以向第三方合作伙伴提供功能。这种做法的思路是，如果你公开了一个设计良好的通用 API，那么你将拥有构建任何类型的应用程序所需的一切。其通用架构如图 4-1 所示。

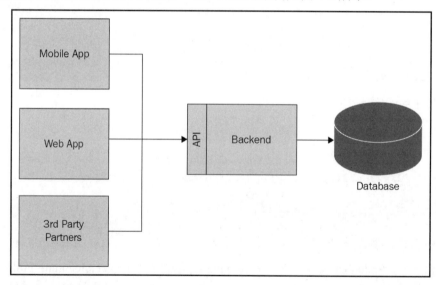

图 4-1

原文	译文	原文	译文
Mobile App	移动 App	Backend	后端
Web App	Web 应用程序	Database	数据库
3rd Party Partners	第三方合作伙伴		

上述方法的缺陷在于,它假定第一方(移动 App 和 Web 应用程序)和第三方(合作伙伴)应用程序的需求始终保持一致,而这种情况很少发生。更常见的是,你希望第三方集成中的某些功能类型和第一方客户端中的功能集是不一样的。

此外,你还希望容忍(甚至是鼓励)第一方客户端中的更改——你的客户端应用程序将不断发展,并且会不断更改其 API 要求。

最后,你无法预期第三方合作伙伴使用你的 API 实现的所有可能用例,因此通用设计是有益的,但是这需要你能够预期移动 App 和 Web 应用程序的需求,而且过于笼统的 API 设计通常也会阻碍你的产品的需求。一个很好的例子是一个服务器端网站,该网站可被重写为单页 JavaScript 应用程序。但在使用通用 API 的情况下,此类项目可能会导致需要数十个 XMLHttpRequest 的页面视图来呈现单个页面视图。

服务于前端的后端(Backend For Frontend,BFF)是一种架构模式,涉及为客户端应用程序的不同类创建单独的、定制的 API。根据你要支持的客户端应用程序的类别多少,你可以开发单独的 BFF 层,而不是在架构中仅有单个 API 层。如何对客户端进行分类完全取决于你的业务需求。开发人员可以决定为所有移动客户端使用单个 BFF 层,也可以将它们分为 iOS BFF 和 Android BFF。同样,你也可以选择为 Web 应用程序和第三方合作伙伴提供单独的 BFF 层,如图 4-2 所示。

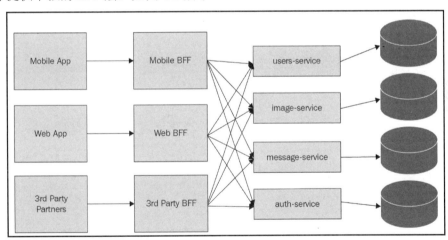

图 4-2

原　　文	译　　文	原　　文	译　　文
Mobile App	移动 App	Mobile BFF	移动 BFF
Web App	Web 应用程序	3rd Party BFF	为第三方合作伙伴提供的 BFF
3rd Party Partners	第三方合作伙伴		

在上述系统中，每种类别的客户端都向自己的 BFF 层发出请求，然后 BFF 层可以聚合对下游服务的调用，并构建一个整体的、定制的 API。

4.3.2 实战操作

为了设计和构建 BFF 层，我们应该首先设计 API。实际上，我们已经在这样做了。在前面的秘笈中，我们演示了使用 CompletableFuture 以异步方式向我们的系统和 social-service 服务发出请求，然后对于返回的每个用户，向 user-details-service 发出异步请求以获取某些用户的配置文件信息。对于我们的移动 App 来说，这是 BFF 层的绝佳用例。想象一下，我们的移动 App 具有一个屏幕，其中显示了该用户关注的用户列表，包括他们的基本信息，例如其头像、用户名和显示名称等。由于社交图信息（用户已经关注的用户列表）和用户个人资料信息（头像、用户名和显示名称）是两项单独服务的职责，因此，如果按照以前的做法的话，那么这是很麻烦的，它要求我们的移动客户端聚合对这些服务的调用，以显示已关注好友的页面。相反，我们可以创建一个移动 BFF 层来处理此聚合，并向客户端返回一个很方便的响应。我们的请求端点如下：

```
GET /users/:user_id/following
```

我们希望得到的响应主体如下：

```
{
    "username": "paulosman",
    "followings": [
        {
            "username": "friendlyuser",
            "display_name": "Friendly User",
            "avatar_url": "http://example.com/pic.jpg"
        },
        {
            ...
        }
    ]
}
```

如我们所见，BFF 将返回一个响应，其中包含在移动 App 中显示已关注好友的屏幕时所需的所有信息。

（1）创建一个名为 bff-mobile 的新 Gradle/Java 项目，其 build.gradle 文件将包含以下内容：

```
group 'com.packtpub.microservices'
version '1.0-SNAPSHOT'

buildscript {
    repositories {
        mavenCentral()
    }
    dependencies {
        classpath group: 'org.springframework.boot',
        name: 'spring-boot-gradle-plugin',
        version: '1.5.9.RELEASE'
    }
}

apply plugin: 'java'
apply plugin: 'org.springframework.boot'

sourceCompatibility = 1.8

repositories {
    mavenCentral()
}

dependencies {
    testCompile group: 'junit', name: 'junit', version: '4.12'
    compile group: 'org.springframework.boot',
    name: 'spring-boot-starter-web'
}
```

（2）创建一个名为 com.packtpub.microservices.ch04.mobilebff 的新程序包和一个名为 Main 的新类。具体如下：

```
package com.packtpub.microservices.ch04.mobilebff;

import org.springframework.boot.SpringApplication;
import org.springframework.boot.autoconfigure.SpringBootApplication;

@SpringBootApplication
public class Main {
    public static void main(String[] args) {
        SpringApplication.run(Main.class, args);
    }
}
```

（3）创建一个名为 com.packtpub.microservices.ch04.mobilebff.models 的新程序包和一个名为 User 的新类。具体如下：

```java
package com.packtpub.microservices.ch04.mobilebff.models;

import com.fasterxml.jackson.annotation.JsonProperty;

public class User {
    private String username;

    @JsonProperty("display_name")
    private String displayName;

    @JsonProperty("avatar_url")
    private String avatarUrl;

    public User() {}

    public User(String username, String displayName,
    String avatarUrl) {
        this.username = username;
        this.displayName = displayName;
        this.avatarUrl = avatarUrl;
    }

    public String getUsername() {
        return username;
    }

    public void setUsername(String username) {
        this.username = username;
    }

    public String getDisplayName() {
        return displayName;
    }

    public void setDisplayName(String displayName) {
        this.displayName = displayName;
    }

    public String getAvatarUrl() {
```

```java
        return avatarUrl;
    }

    public void setAvatarUrl(String avatarUrl) {
        this.avatarUrl = avatarUrl;
    }

    public String toString() {
        return String.format(
                "[User username:%s, displayName:%s, avatarUrl:%s]",
                username, displayName, avatarUrl);
    }
}
```

(4)创建另一个模块,将其命名为 **Followings**。具体如下:

```java
package com.packtpub.microservices.ch04.mobilebff.models;

import java.util.List;

public class Followings {
    private String username;

    private List<String> followings;

    public Followings() {}

    public Followings(String username, List<String> followings) {
        this.username = username;
        this.followings = followings;
    }

    public String getUsername() {
        return username;
    }

    public void setUsername(String username) {
        this.username = username;
    }

    public List<String> getFollowings() {
        return followings;
    }
```

```
    public void setFollowings(List<String> followings) {
        this.followings = followings;
    }
}
```

（5）我们将创建的最后一个模块称为 HydratedFollowings。它和 Followings 模块比较相似，但不是将用户列表存储为字符串，而是包含 User 对象的列表。具体如下：

```
package com.packtpub.microservices.ch04.mobilebff.models;

import java.util.List;

public class HydratedFollowings {
    private String username;

    private List<User> followings;

    public HydratedFollowings() {}

    public HydratedFollowings(String username, List<User> followings) {
        this.username = username;
        this.followings = followings;
    }

    public String getUsername() {
        return username;
    }

    public void setUsername(String username) {
        this.username = username;
    }

    public List<User> getFollowings() {
        return followings;
    }

    public void setFollowings(List<User> followings) {
        this.followings = followings;
    }
}
```

(6)创建服务客户端。创建一个名为 com.packtpub.microservices.ch04.mobilebff.services 的新程序包和一个名为 SocialGraphService 的新类。具体如下:

```java
package com.packtpub.microservices.ch04.mobilebff.services;

import com.packtpub.microservices.ch04.mobilebff.models.Followings;
import org.springframework.boot.web.client.RestTemplateBuilder;
import org.springframework.scheduling.annotation.Async;
import org.springframework.stereotype.Service;
import org.springframework.web.client.RestTemplate;

import java.util.concurrent.CompletableFuture;

@Service
public class SocialGraphService {

    private final RestTemplate restTemplate;

    public SocialGraphService(RestTemplateBuilder
    restTemplateBuilder) {
        this.restTemplate = restTemplateBuilder.build();
    }

    @Async
    public CompletableFuture<Followings>
    getFollowing(String username) {
        String url =
String.format("http://localhost:4567/followings/
        %s", username);
        Followings followings = restTemplate.getForObject(url,
        Followings.class);
        return CompletableFuture.completedFuture(followings);
    }
}
```

(7)创建一个名为 UsersService 的新类,它将作为我们的用户服务的客户端。具体如下:

```java
package com.packtpub.microservices.ch04.mobilebff.services;

import com.packtpub.microservices.ch04.mobilebff.models.User;
import org.springframework.boot.web.client.RestTemplateBuilder;
import org.springframework.scheduling.annotation.Async;
```

```
import org.springframework.stereotype.Service;
import org.springframework.web.client.RestTemplate;

import java.util.concurrent.CompletableFuture;

@Service
public class UsersService {

    private final RestTemplate restTemplate;

    public UsersService(RestTemplateBuilder restTemplateBuilder) {
        this.restTemplate = restTemplateBuilder.build();
    }

    @Async
    public CompletableFuture<User> getUserDetails(String username)
{
        String url = String.format("http://localhost:4568/users/
        %s", username);
        User user = restTemplate.getForObject(url, User.class);
        return CompletableFuture.completedFuture(user);
    }
}
```

（8）现在可以通过创建公开端点的控制器将步骤（2）～（7）中创建的包、类和模块绑定在一起。如果你已经完成了前面的秘笈，那么下面这段代码看起来会很熟悉，因为我们使用了完全相同的模式来对依赖性的异步服务调用进行建模。

创建一个名为 com.packtpub.microservices.ch04.mobilebff.controllers 的程序包和一个名为 UsersController 的新类：

```
package com.packtpub.microservices.ch04.mobilebff.controllers;

import
com.packtpub.microservices.ch04.mobilebff.models.HydratedFollowings;
import com.packtpub.microservices.ch04.mobilebff.models.User;
import
com.packtpub.microservices.ch04.mobilebff.services.SocialGraphService;
import com.packtpub.microservices.ch04.mobilebff.services.UsersService;
import org.springframework.beans.factory.annotation.Autowired;
import org.springframework.web.bind.annotation.*;

import java.util.List;
```

```java
import java.util.concurrent.CompletableFuture;
import java.util.concurrent.ExecutionException;
import java.util.stream.Collectors;

@RestController
public class UsersController {

    @Autowired
    private SocialGraphService socialGraphService;

    @Autowired
    private UsersService userService;

    @RequestMapping(path = "/users/{username}/followings",
    method = RequestMethod.GET)

    public HydratedFollowings getFollowings(@PathVariable String username)
    throws ExecutionException, InterruptedException {
        CompletableFuture<List<User>> users =
socialGraphService.getFollowing
        (username).thenApply(f -> f.getFollowings().stream().map(
                    u -> userService.getUserDetails(u)).map(
CompletableFuture::join).collect(Collectors.toList()));
        return new HydratedFollowings(username, users.get());
    }
}
```

（9）在编写完成之后，运行该应用程序并发送 GET 请求到/users/username/followings 中。此时，你应该返回一个完全混合的 JSON 响应，其中包含用户的用户名和该用户已经关注的每个用户的详细信息。

4.4 使用 JSON 和 HTTP 实现 RPC 一致性

构建多个微服务时，服务之间的一致性和约定开始产生实际影响。当微服务架构中出现问题时，你可能最终需要花大量的时间来调试这些服务。因此，如果能够对特定服务接口的性质做出某些假设，则可以节省大量的时间和精力。

4.4.1 理论阐释

拥有一致的远程过程调用（Remote Procedure Call，RPC）方法，可以使你将某些问题整理编写到可以在服务之间轻松共享的库中。通过采用一致的方法，可以简化诸如身份验证、应如何解释标头、响应主体中包含哪些信息，以及如何请求分页响应之类的事情。此外，报告错误的方式也应尽可能保持一致。

由于微服务架构通常由不同的团队以不同的编程语言编写的服务组成，因此必须使用一切方法努力实现一致的 RPC 语义，这可能是以库的方式，使用与构建服务时一样多的语言来实现。这可能会很麻烦，但为了客户端与各种服务进行对话时可以假定一致性，这样的麻烦是值得的。

在本秘笈中，我们将重点介绍在 Java 中使用 Spring Boot 编写的服务。我们将编写一个自定义的序列化器（Serializer），以一致的方式显示资源和资源集合，其中包括分页信息。然后，我们将修改 message-service 服务以使用新的序列化器。

4.4.2 实战操作

在本秘笈中，我们将创建一个包装器（Wrapper）类来表示带有分页信息的资源集合。我们还将使用来自 jackson 库的 JsonRootName 注解使单个资源表示形式保持一致。

首先可以修改前面秘笈中介绍过的 message-service 服务，将以下代码添加到其中。

（1）创建一个名为 ResourceCollection 的新类。此类将是一个普通的简单 Java 对象（Plain Ordinary Java Object，POJO），其中包含代表页码的字段、项目列表，以及可用于访问集合中下一页的 URL。具体如下：

```java
package com.packtpub.microservices.ch04.message.models;

import com.fasterxml.jackson.annotation.JsonProperty;
import com.fasterxml.jackson.annotation.JsonRootName;

import java.util.List;

@JsonRootName("result")
public class ResourceCollection<T> {

    private int page;

    @JsonProperty("next_url")
```

```
    private String nextUrl;

    private List<T> items;

    public ResourceCollection(List<T> items, int page, String
nextUrl) {
        this.items = items;
        this.page = page;
        this.nextUrl = nextUrl;
    }

    public int getPage() {
        return page;
    }

    public void setPage(int pageNumber) {
        this.page = page;
    }

    public String getNextUrl() {
        return nextUrl;
    }

    public void setNextUrl(String nextUrl) {
        this.nextUrl = nextUrl;
    }

    public List<T> getItems() {
        return items;
    }

    public void setItems(List<T> items) {
        this.items = items;
    }
}
```

（2）创建或修改 Message 模块。在这里可以使用 JsonRootName 注解将 Message 表示形式包装在单个 JSON 对象（该对象还包括 item 键）中。为了获得一致性的表示，还应该将它们添加到我们的服务作为资源公开的所有模块中。具体如下：

```
package com.packtpub.microservices.ch04.message.models;

import com.fasterxml.jackson.annotation.JsonRootName;
```

```java
@JsonRootName("item")
public class Message {
    private String id;
    private String toUser;
    private String fromUser;
    private String body;

    public Message(String id, String toUser, String fromUser, String body) {
        this.id = id;
        this.toUser = toUser;
        this.fromUser = fromUser;
        this.body = body;
    }

    public String getId() {
        return id;
    }

    public void setId(String id) {
        this.id = id;
    }

    public String getToUser() {
        return toUser;
    }

    public void setToUser(String toUser) {
        this.toUser = toUser;
    }

    public String getFromUser() {
        return fromUser;
    }

    public void setFromUser(String fromUser) {
        this.fromUser = fromUser;
    }

    public String getBody() {
        return body;
```

```
    }

    public void setBody(String body) {
        this.body = body;
    }
}
```

（3）以下控制器将返回一个消息列表和一条特定消息。我们将消息列表包装在先前创建的 ResourceCollection 类中：

```
package com.packtpub.microservices.ch04.message.controllers;

import com.packtpub.microservices.ch04.message.models.Message;
import com.packtpub.microservices.ch04.message.models.ResourceCollection;
import org.springframework.web.bind.annotation.*;

import javax.servlet.http.HttpServletRequest;
import java.util.List;
import java.util.stream.Collectors;
import java.util.stream.Stream;

@RestController
public class MessageController {

    @RequestMapping(value = "/messages", method = RequestMethod.GET)
    public ResourceCollection<Message> messages(@RequestParam(name="page", required=false, defaultValue="1") int page,
                                                 HttpServletRequest request)
    {
        List<Message> messages = Stream.of(
                new Message("1234","paul", "veronica", "hello!"),
                new Message("5678","meghann", "paul", "hello!")
        ).collect(Collectors.toList());

        String nextUrl = String.format("%s?page=%d", request.getRequestURI(), page + 1);

        return new ResourceCollection<>(messages, page, nextUrl);
    }
```

```
    @RequestMapping(value = "/messages/{id}", method =
RequestMethod.GET)
    public Message message(@PathVariable("id") String id) {
        return new Message(id, "paul", "veronica", "hi dad");
    }
}
```

(4)如果通过对/messages 发出请求来测试请求项目集合,那么现在应该返回以下 JSON:

```
{
    "result": {
        "page": 1,
        "items": [
            {
                "id": "1234",
                "toUser": "paul",
                "fromUser": "veronica",
                "body": "hello!"
            },
            {
                "id": "5678",
                "toUser": "meghann",
                "fromUser": "paul",
                "body": "hello!"
            }
        ],
        "next_url": "/messages?page=2"
    }
}
```

(5)对于单个资源,应该会返回以下 JSON:

```
{
    "item": {
        "id": "123",
        "toUser": "paul",
        "fromUser": "veronica",
        "body": "hi dad"
    }
}
```

对资源或资源列表的表示方式进行一些标准化可以极大地简化微服务架构中服务的处理。但是,使用 JSON 和 HTTP 进行此操作涉及大量的手动工作,因此可以考虑将其抽象化。在接下来的秘笈中，我们将探索使用 Thrift 和 gRPC，这是 RPC 的 HTTP/JSON 表示方式的两种替代方法。

4.5 使用 Thrift

JSON 和 HTTP 是用于数据传输和定义的简单、直接的解决方案，完全可以满足许多微服务架构的目的。但是，如果想要实现类型安全并且需要更好的性能，那么值得考虑诸如 Thrift 或 gRPC 之类的二进制解决方案。

4.5.1 理论阐释

Apache Thrift 是一种由 Facebook 发明的接口定义语言（Interface Definition Language，IDL）和二进制传输协议。它允许你通过定义结构（与大多数语言中的对象相似）和服务公开的异常来指定 API。IDL 中定义的 Thrift 接口可用于以受支持的语言生成代码，然后将其用于管理 RPC 调用。受支持的语言包括 C、C++、Python、Ruby 和 Java。

二进制协议（如 Thrift）的好处主要是提高了性能和类型安全性。根据使用的 JSON 库，对较大的 JSON 有效负载进行序列化和反序列化的成本可能会非常高，并且 JSON 没有客户端可以在处理响应时使用的任何类型系统。另外，由于 Thrift 包含可用于以任何受支持的语言生成代码的 IDL，因此让 Thrift 处理客户端和服务器代码的生成很容易，从而减少了需要完成的手动工作量。

由于 Apache Thrift 不使用 HTTP 作为传输层，因此导出 Thrift 接口的服务将启动其自己的 Thrift 服务器。在本秘笈中，我们将为 message-service 服务定义 IDL，并使用 Thrift 生成处理程序代码。然后，我们将创建服务器样板，它处理启动服务、侦听指定端口等操作。

4.5.2 实战操作

本秘笈需要执行以下操作。

（1）使用以下 build.gradle 文件创建一个新的 Gradle/Java 项目：

```
group 'com.packtpub.microservices'
version '1.0-SNAPSHOT'

buildscript {
    repositories {
```

```
        maven {
            url "https://plugins.gradle.org/m2/"
        }
    }
    dependencies {
        classpath "gradle.plugin.org.jruyi.gradle:thrift-gradle-plugin:0.4.0"
    }
}

apply plugin: 'java'
apply plugin: 'org.jruyi.thrift'
apply plugin: 'application'

mainClassName = 'com.packtpub.microservices.ch04.MessageServiceServer'

compileThrift {
    recurse true

    generator 'html'
    generator 'java', 'private-members'
}

sourceCompatibility = 1.8

repositories {
    mavenCentral()
}

dependencies {
    compile group: 'org.apache.thrift', name: 'libthrift', version: '0.11.0'
    testCompile group: 'junit', name: 'junit', version: '4.12'
}
```

（2）创建一个名为 src/main/thrift 的目录和一个名为 service.thrift 的文件。这是服务的 IDL 文件。我们将定义一个 MessageException 异常、实际的 Message 对象和一个 MessageService 接口。有关 Thrift IDL 文件的特定语法的更多信息可以访问 Thrift 项目网站上提供的说明文档，其网址如下：

https://thrift.apache.org/docs/idl

为简单起见，我们将在服务中定义一个方法，以返回特定用户的消息列表。具体如下：

```
namespace java com.packtpub.microservices.ch04.thrift

exception MessageException {
    1: i32 code,
    2: string description
}

struct Message {
    1: i32 id,
    2: string from_user,
    3: string to_user,
    4: string body
}

service MessageService {
    list<Message> inbox(1: string username) throws
(1:MessageException e)
}
```

（3）运行汇编 Gradle 任务将为先前的 IDL 生成代码。现在，我们将创建 MessageService 类的实现。这将从先前的 IDL 扩展自动生成的接口。为简单起见，我们的 MessageService 实现将不会连接到任何数据库，而是使用收件箱（Inboxe）的静态的、硬编码的表示形式（收件箱将在构造函数中被构建）。

```
package com.packtpub.microservices.ch04.thrift;

import com.packtpub.microservices.ch04.thrift.Message;
import com.packtpub.microservices.ch04.thrift.MessageException;
import com.packtpub.microservices.ch04.thrift.MessageService;
import org.apache.thrift.TException;

import java.util.HashMap;
import java.util.List;
import java.util.Map;
import java.util.stream.Collectors;
import java.util.stream.Stream;

public class MessageServiceImpl implements MessageService.Iface {

    private Map<String, List<Message>> messagesRepository;
```

```
    MessageServiceImpl() {
        // 使用一些示例消息来填充我们的模拟存储库
        messagesRepository = new HashMap<>();
        messagesRepository.put("usertwo", Stream.of(
            new Message(1234, "userone", "usertwo", "hi"),
            new Message(5678, "userthree", "usertwo", "hi")
        ).collect(Collectors.toList()));
        messagesRepository.put("userone", Stream.of(
            new Message(1122, "usertwo", "userone", "hi"),
            new Message(2233, "userthree", "userone", "hi")
        ).collect(Collectors.toList()));
    }

    @Override
    public List<Message> inbox(String username) throws TException {
        if (!messagesRepository.containsKey(username))
            throw new MessageException(100, "Inbox is empty");
        return messagesRepository.get(username);
    }
}
```

（4）创建服务器。创建一个名为 MessageServiceServer 的新类，具体如下：

```
package com.packtpub.microservices.ch04.thrift;

import com.packtpub.microservices.ch04.thrift.MessageService;
import org.apache.thrift.server.TServer;
import org.apache.thrift.server.TSimpleServer;
import org.apache.thrift.transport.TServerSocket;
import org.apache.thrift.transport.TServerTransport;
import org.apache.thrift.transport.TTransportException;

public class MessageServiceServer {

    private TSimpleServer server;

    private void start() throws TTransportException {
        TServerTransport serverTransport = new TServerSocket(9999);
        server = new TSimpleServer(new TServer.Args(serverTransport)
            .processor(new MessageService.Processor<>(new MessageServiceImpl())));
        server.serve();
```

```
    }

    private void stop() {
        if (server != null && server.isServing())
            server.stop();
    }

    public static void main(String[] args) {
        MessageServiceServer service = new MessageServiceServer();
        try {
            if (args[1].equals("start"))
                service.start();
            else if (args[2].equals("stop"))
                service.stop();
        } catch (TTransportException e) {
            e.printStackTrace();
        }
    }
}
```

服务现已构建完毕,并且已经将 Apache Thrift 应用于 RPC。你可以进行下一步的练习,尝试使用相同的 IDL 生成可用于调用此服务的客户端代码。

4.6 使用 gRPC

4.5 节介绍了 Thrift,本节将介绍同为二进制解决方案的 gRPC。

4.6.1 理论阐释

gRPC 最初是由 Google 发明的远程过程调用(RPC)框架。与 Thrift 不同,gRPC 利用了现有技术,特别是协议缓冲区(Protocol Buffer),用于其 IDL,并且将 HTTP/2 用于其传输层。在完成了前面的秘笈之后,你应该会感觉到 gRPC 与 Thrift 在很多方面都很相似。不过,与 Thrift IDL 不同的是,其类型和服务是在.proto 文件中定义的,并且.proto 文件可以通过协议缓冲区的编译器生成代码。

4.6.2 实战操作

本秘笈需要执行以下操作。

（1）使用以下 build.gradle 文件创建一个新的 Gradle/Java 项目。值得一提的是，由于需要安装和配置 protobuf Gradle 插件，才能使用 Gradle 从 protobuf 文件生成代码。因此，这里需要将 protobuf 库列为依赖项。最后，还必须告诉集成开发环境（IDE）到哪里去寻找已生成的类。具体如下：

```
group 'com.packtpub.microservices'
version '1.0-SNAPSHOT'

buildscript {
    repositories {
        mavenCentral()
    }
    dependencies {
        classpath 'com.google.protobuf:protobuf-gradle-plugin:0.8.3'
    }
}

apply plugin: 'java'
apply plugin: 'com.google.protobuf'
apply plugin: 'application'

mainClassName = 'com.packtpub.microservices.ch04.grpc.MessageServer'

sourceCompatibility = 1.8

repositories {
    mavenCentral()
}

def grpcVersion = '1.10.0'

dependencies {
    compile group: 'com.google.api.grpc', name: 'proto-google-common-protos', version: '1.0.0'
    compile group: 'io.grpc', name: 'grpc-netty', version: grpcVersion
    compile group: 'io.grpc', name: 'grpc-protobuf', version: grpcVersion
    compile group: 'io.grpc', name: 'grpc-stub', version: grpcVersion
```

```
        testCompile group: 'junit', name: 'junit', version: '4.12'
}

protobuf {
    protoc {
        artifact = 'com.google.protobuf:protoc:3.5.1-1'
    }
    plugins {
        grpc {
            artifact = "io.grpc:protoc-gen-grpc-java:${grpcVersion}"
        }
    }
    generateProtoTasks {
        all()*.plugins {
            grpc {}
        }
    }
}

// 向 IntelliJ IDEA、Eclipse 或 NetBeans 等 IDE 告知已生成的代码
sourceSets {
    main {
        java {
            srcDirs 'build/generated/source/proto/main/grpc'
            srcDirs 'build/generated/source/proto/main/java'
        }
    }
}
```

（2）创建一个名为 src/main/proto 的新目录和一个名为 message_service.proto 的新文件。这就是为我们的服务提供的 protobuf 的定义。和上一个秘笈一样，为简单起见，我们将仅公开一个方法，该方法将返回指定用户的消息列表。具体如下：

```
option java_package = "com.packtpub.microservices.ch04.grpc";

message Username {
    required string username = 1;
}

message Message {
    required string id = 1;
    required string from_user = 2;
```

```
    required string to_user = 3;
    required string body = 4;
}

message InboxReply {
    repeated Message messages = 1;
}

service MessageService {
    rpc inbox(Username) returns (InboxReply) {}
}
```

(3) 实现实际的服务。为此，我们需要创建一个名为 MessageServer 的新类，其中包含用于启动和停止服务器的所有必需样板。我们还将创建一个名为 MessageService 的内部类，该类扩展了已生成的 MessageServiceGrpc.MessageServiceImplBase 类。具体如下：

```
package com.packtpub.microservices.ch04.grpc;

import io.grpc.Server;
import io.grpc.ServerBuilder;
import io.grpc.stub.StreamObserver;

import java.io.IOException;

public class MessageServer {

    private final int port;
    private final Server server;

    private MessageServer(int port) throws IOException {
        this(ServerBuilder.forPort(port), port);
    }

    private MessageServer(ServerBuilder<?> serverBuilder, int port)
{
        this.port = port;
        this.server = serverBuilder.addService(new
MessageService()).build();
    }

    public void start() throws IOException {
        server.start();
```

```java
        Runtime.getRuntime().addShutdownHook(new Thread() {
            @Override
            public void run() {
                // 在这里使用 stderr,因为日志管理程序可能已被其 Java 虚拟机(JVM)
                // 关闭钩子重置
                System.err.println("*** shutting down gRPC server since JVM is shutting down");
                MessageServer.this.stop();
                System.err.println("*** server shut down");
            }
        });
    }

    public void stop() {
        if (server != null) {
            server.shutdown();
        }
    }

    private void blockUntilShutdown() throws InterruptedException {
        if (server != null) {
            server.awaitTermination();
        }
    }

    private static class MessageService extends MessageServiceGrpc.MessageServiceImplBase {
        public void inbox(MessageServiceOuterClass.Username request, StreamObserver<MessageServiceOuterClass.InboxReply> responseObserver) {
            MessageServiceOuterClass.InboxReply reply = MessageServiceOuterClass.InboxReply.newBuilder().addMessages(
                MessageServiceOuterClass.Message.newBuilder()
                    .setId("1234")
                    .setFromUser("Paul")
                    .setToUser("Veronica")
                    .setBody("hi")
            ).addMessages(
                MessageServiceOuterClass.Message.newBuilder()
                    .setId("5678")
                    .setFromUser("FooBarUser")
```

```
                        .setToUser("Veronica")
                        .setBody("Hello again")
                ).build();
                responseObserver.onNext(reply);
                responseObserver.onCompleted();
            }
        }

        public static void main(String[] args) throws Exception {
            MessageServer server = new MessageServer(8989);
            server.start();
            server.blockUntilShutdown();
        }
    }
```

第 5 章　可靠性模式

本章包含以下操作秘笈。
- 使用断路器实现背压。
- 使用指数退避算法重试请求。
- 通过缓存提高性能。
- 通过 CDN 提供更高效的服务。
- 优雅地降低用户体验。
- 通过游戏日演习验证容错能力。
- 引入自动化混沌工程。

5.1　导　　语

在分布式系统领域中，可靠性正成为越来越流行的话题。搜索人才市场，可以看到站点可靠性工程师（Site Reliability Engineer，SRE）或混沌工程师（Chaos Engineer）的职位发布正变得越来越常见，并且随着越来越多的组织朝着使用云原生（Cloud-Native）技术的方向发展，不可忽视的现实是系统故障始终存在。网络将会遇到拥塞，交换机或其他硬件组件将会发生故障，并且系统中的大量潜在故障模式将使生产环境中的客服人员应接不暇。完全防止故障是不可能的，因此我们应该尝试将系统设计为尽可能容忍故障。

微服务为可靠性设计提供了有趣而有用的机会。因为微服务鼓励开发人员将系统分解为封装单一职责的服务，所以可以使用许多有用的可靠性模式来隔离发生故障的情况。在规划可靠性时，微服务架构也提出了许多挑战。例如，它对网络请求、异构配置、多个数据存储和连接池以及不同技术堆栈的依赖性越来越高，这一切都导致固有的更加复杂的环境，其中可以出现不同样式的故障模式。

无论是处理微服务架构还是单体架构代码库，都需要从根本上解决系统在某种故障状态下正常运行的问题。可以访问以下链接以获取更多信息：

https://www.youtube.com/watch?v=tZ2wj2pxO6Q

从一开始就在系统中建立弹性，这使我们能够优化在上述故障情况下的反应方式。

在本章中，我们将讨论许多有用的可靠性模式，这些可靠性模式可以在设计和构建微服务时使用，以减少预期和意外的系统故障的影响。

5.2 使用断路器实现背压

分布式系统中的故障可能很难调试，因为其表象（如延迟尖峰或高错误率）看起来可能与根本原因（例如数据库查询速度慢、垃圾回收周期导致服务减慢了请求的处理速度等）相去甚远。

有时，系统的完全中断只是其中一小部分发生故障的结果，尤其是当系统的组件难以处理增加的负载时。

因此，只要有可能，我们都希望防止系统某个部分的故障级联传导到其他部分，从而导致广泛且难以调试的生产问题。此外，如果故障是暂时的，那么我们希望系统能够在故障结束后自我修复。如果某个特定服务由于负载的临时峰值而出现问题，那么我们在设计系统时应该考虑到这种情况，防止对不正常的服务发出请求，以便让这种不正常的服务有时间进行恢复，在该服务回归正常之后再对它发出请求。

5.2.1 理论阐释

在日常生活中，每个家庭的电闸都使用了断路器，以防止过度用电加热内部布线并烧毁房屋。如果断路器检测到电路过载，则电路将跳闸，并且无法处理从电路中汲取电流。经过一段时间后，断路器可以再次闭合，以使系统正常运行。

相同的方法也可以转换为软件，并应用于微服务架构。当一个服务调用另一个服务时，我们应该将 RPC 调用包装在断路器中。如果请求反复失败，表明该服务不正常，则断路器将断开，从而阻止尝试其他请求。然后，调用服务可以"快速失败"并决定如何处理失败模式。经过一段时间之后（该时间的长短可配置调整），我们可以允许另一个请求通过，如果成功，则再次闭合电路，使系统恢复正常运行。相关流程图如图 5-1 所示。

实现断路器的库可用于大多数流行的编程语言。由 Netflix 建立的 Hystrix 容错库（在前面的秘笈中我们已经使用过它）就是这样一个库。Twitter 的 Finagle 等某些框架会自动将 RPC 包装在断路器中，以跟踪故障并自动管理断路器的状态。开源服务网格软件（如 Conduit 和 Linkerd）也会自动将断路器添加到 RPC 中。在本秘笈中，我们将介绍一个名为 resilience4j 的库，并使用其断路器实现在达到故障阈值时允许从一项服务到另一项服务的调用快速失败。为了使示例更具体，我们将修改 message-service 服务，该服务将调用 social-graph-service 服务以确定两个用户是否已经彼此关注，并将 RPC 调用包装在断路器中。

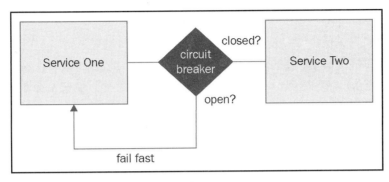

图 5-1

原　　文	译　　文	原　　文	译　　文
Service One	服务 1	open?	断开？
circuit breaker	断路器	fail fast	快速失败
closed?	闭合？		

5.2.2　实战操作

为了演示断路器中的包装服务调用，我们将创建 pichat 程序 message-service 服务的版本，该版本将公开用于发送和检索消息的端点。要将消息从发送者发送给接收者，这两个用户必须已经相互关注。这种好友关系由 social-graph-service 服务处理。为简单起见，我们将在 Ruby 中编写一个简单的模拟 social-graph-service 服务，就像在以前的秘笈中所做的那样。该模拟服务将公开一个端点，该端点列出了指定用户的好友关系。以下就是 Ruby 中的模拟 social-graph-service 服务的源代码：

```
require 'sinatra'
require 'json'

get '/friendships/:username' do
    content_type :json
    {
        'username': params[:username],
        'friendships': [
            'pichat:users:johndoe',
            'pichat:users:janesmith',
```

```
            'pichat:users:anotheruser'
        ]
    }.to_json
end
```

> **注意：**
> 在我们的模拟服务中，使用了 pichat:users:username 格式的字符串来标识系统中的用户。这些是伪 URI，它可以唯一标识我们系统中的用户。现在，你只需要知道它们是用于标识系统中用户的唯一字符串。

我们的模拟 social-graph-service 服务公开了以下单个端点：

```
GET /friendships/paulosman
```

上述端点将返回一个 JSON 响应主体，该主体表示请求的用户拥有的好友。具体如下：

```
{
    "username": "fdsa",
    "friendships": [
        "pichat:users:foobar",
        "pichat:users:asomefdsa"
    ]
}
```

在本地主机 4567 端口（Ruby Sinatra 应用程序的默认端口）上运行模拟的 social-graph-service 服务之后，我们就可以开始编写 message-service 服务了。与以前的秘笈一样，我们将使用 Java 和 Spring Boot 框架。除此之外，我们还将使用 resilience4j 断路器库来包装从 message-service 服务到 social-graph-service 服务的调用。

首先，我们将开发 message-service 服务代码，然后将其添加到 resilience4j 断路器库中，以便为我们的服务添加一定程度的弹性，具体操作步骤如下：

（1）创建一个新的 Gradle/Java 项目，并将以下代码添加到 build.gradle 文件中：

```
group 'com.packtpub.microservices'
version '1.0-SNAPSHOT'

buildscript {
    repositories {
        mavenCentral()
    }
    dependencies {
        classpath group: 'org.springframework.boot', name: 'spring-boot-gradle-plugin', version: '1.5.9.RELEASE'
```

```
        }
}

apply plugin: 'java'
apply plugin: 'org.springframework.boot'

sourceCompatibility = 1.8

repositories {
    mavenCentral()
}

dependencies {
    testCompile group: 'junit', name: 'junit', version: '4.12'
    compile group: 'org.springframework.boot', name: 'spring-boot-starter-web'
}
```

（2）我们的 message-service 服务代码将具有两个自动连接到控制器的 bean。第一个是内存中的消息存储库（在实际示例中，它将被数据库之类的持久保存层取代）；第二个是 social-graph-service 服务的客户端。在创建这些对象之前，还需要创建一些支持对象。所此，创建一个名为 com.packtpub.microservices.ch05.message.exceptions 的新程序包和一个名为 MessageNotFoundException 的新类。它们将用于指示找不到消息，这将导致我们的服务产生 404 响应，具体如下：

```
package com.packtpub.microservices.ch05.message.exceptions;

import org.springframework.http.HttpStatus;
import org.springframework.web.bind.annotation.ResponseStatus;

@ResponseStatus(HttpStatus.NOT_FOUND)
public class MessageNotFoundException extends Exception {
    public MessageNotFoundException(String message) {
super(message); }
}
```

（3）在异常包中创建另一个类，名为 MessageSendForbiddenException。这将用于指示由于发送者和接收者不是好友而无法发送消息。我们的服务对于禁止发送消息的响应代码为 403，具体如下：

```
package com.packtpub.microservices.ch05.message.exceptions;
```

```
import org.springframework.http.HttpStatus;
import org.springframework.web.bind.annotation.ResponseStatus;

@ResponseStatus(HttpStatus.FORBIDDEN)
public class MessageSendForbiddenException extends Exception {
    public MessageSendForbiddenException(String message) {
super(message); }
}
```

(4)创建 SocialGraphClient 类。这需要创建一个名为 com.packtpub.microservices.ch05.message.clients 的新程序包和一个名为 SocialGraphClient 的新类,具体如下:

```
package com.packtpub.microservices.ch05.message.clients;

import com.packtpub.microservices.ch05.models.Friendships;
import org.springframework.web.client.RestTemplate;

import java.util.List;

public class SocialGraphClient {
    private String baseUrl;

    public SocialGraphClient(String baseUrl) {
        this.baseUrl = baseUrl;
    }

    public List<String> getFriendships(String username) {
        String requestUrl = baseUrl + "/friendships/" + username;
        RestTemplate template = new RestTemplate();
        UserFriendships friendships =
template.getForObject(requestUrl, UserFriendships.class);
        return friendships.getFriendships();
    }
}
```

(5)现在可以来创建模块。我们需要一个模块来表示特定用户所拥有的 UserFriendship(好友关系),另外还需要一个模块来表示消息。创建一个名为 com.packtpub. microservices.ch05.message.models 的新程序包和一个名为 Friendships 的新类,具体如下:

```
package com.packtpub.microservices.ch05.message.models;

import java.util.List;
```

```java
public class Friendships {
    private String username;
    private List<String> friendships;

    public Friendships() {
        this.friendships = new ArrayList<>();
    }

    public Friendships(String username) {
        this.username = username;
        this.friendships = new ArrayList<>();
    }

    public Friendships(String username, List<String> friendships) {
        this.username = username;
        this.friendships = friendships;
    }

    public String getUsername() {
        return username;
    }

    public void setUsername(String username) {
        this.username = username;
    }

    public List<String> getFriendships() {
        return friendships;
    }

    public void setFriendships(List<String> friendships) {
        this.friendships = friendships;
    }
}
```

（6）在com.packtpub.microservices.ch05.message.models程序包中创建一个名为Message的新类，具体如下：

```java
package com.packtpub.microservices.ch05.message.models;

import com.fasterxml.jackson.annotation.JsonProperty;

public class Message {
```

```java
    private String id;
    private String sender;
    private String recipient;
    private String body;
    @JsonProperty("attachment_uri")
    private String attachmentUri;

    public Message() {}

    public Message(String sender, String recipient, String body, String attachmentUri) {
        this.sender = sender;
        this.recipient = recipient;
        this.body = body;
        this.attachmentUri = attachmentUri;
    }

    public Message(String id, String sender, String recipient, String body, String attachmentUri) {
        this.id = id;
        this.sender = sender;
        this.recipient = recipient;
        this.body = body;
        this.attachmentUri = attachmentUri;
    }

    public String getId() {
        return id;
    }

    public String getSender() {
        return sender;
    }

    public void setSender(String sender) {
        this.sender = sender;
    }

    public String getRecipient() {
        return recipient;
    }
```

```
    public void setRecipient(String recipient) {
        this.recipient = recipient;
    }

    public String getBody() {
        return body;
    }

    public void setBody(String body) {
        this.body = body;
    }

    public String getAttachmentUri() {
        return attachmentUri;
    }

    public void setAttachmentUri(String attachmentUri) {
        this.attachmentUri = attachmentUri;
    }
}
```

（7）在创建模块之后，即可转到内存中的消息存储库中。该类仅使用 HashMap 来存储由 UUID 键提供的消息。这些消息不是被持久保存的，在重新启动服务后将无法存在，因此对于生产服务来说，建议不要使用此技术。该类有两个方法：一个是 saved 方法，它将生成 UUID 并将消息存储在映射中；另一个是 get 方法，它将尝试从映射中检索消息。如果未找到任何消息，那么将引发异常。具体如下：

```
package com.packtpub.microservices.ch05.message;

import com.packtpub.microservices.ch05.message.exceptions.MessageNotFoundException;
import com.packtpub.microservices.ch05.message.models.Message;

import java.util.HashMap;
import java.util.Map;
import java.util.UUID;

public class MessageRepository {

    private Map<String, Message> messages;

    public MessageRepository() {
```

```
        messages = new HashMap<>();
    }

    public Message save(Message message) {
        UUID uuid = UUID.randomUUID();
        Message saved = new Message(uuid.toString(),
message.getSender(), message.getRecipient(),
                message.getBody(), message.getAttachmentUri());
        messages.put(uuid.toString(), saved);
        return saved;
    }

    public Message get(String id) throws MessageNotFoundException {
        if (messages.containsKey(id)) {
            Message message = messages.get(id);
            return message;
        } else {
            throw new MessageNotFoundException("Message " + id + " could not be found");
        }
    }
}
```

（8）我们的服务具有单个消息控制器。该控制器有两个端点：一个端点允许调用方按 ID 检索消息（如果找不到消息，则返回 404 响应）；另一个端点则会尝试发送消息（如果消息的发送者和接收者不是好友，则返回 403 响应）。具体如下：

```
package com.packtpub.microservices.ch05.message;

import com.packtpub.microservices.ch05.message.clients.SocialGraphClient;
import com.packtpub.microservices.ch05.message.exceptions.
MessageNotFoundException;
import com.packtpub.microservices.ch05.message.exceptions.
MessageSendForbiddenException;
import com.packtpub.microservices.ch05.message.models.Message;
import org.springframework.beans.factory.annotation.Autowired;
import org.springframework.http.ResponseEntity;
import org.springframework.web.bind.annotation.*;
import org.springframework.web.servlet.support.
ServletUriComponentsBuilder;

import java.net.URI;
import java.util.List;
```

```
@RestController
public class MessageController {

    @Autowired
    private MessageRepository messagesStore;

    @Autowired
    private SocialGraphClient socialGraphClient;

    @RequestMapping(path = "/messages/{id}", method =
RequestMethod.GET, produces = "application/json")
    public Message get(@PathVariable("id") String id) throws
MessageNotFoundException {
        return messagesStore.get(id);
    }

    @RequestMapping(path = "/messages", method =
RequestMethod.POST, produces = "application/json")
    public ResponseEntity<Message> send(@RequestBody Message
message) throws MessageSendForbiddenException {

        List<String> friendships =
socialGraphClient.getFriendships(message.getSender());
        if (!friendships.contains(message.getRecipient())) {
            throw new MessageSendForbiddenException("Must be
friends to send message");
        }

        Message saved = messagesStore.save(message);
        URI location = ServletUriComponentsBuilder
            .fromCurrentRequest().path("/{id}")
            .buildAndExpand(saved.getId()).toUri();
        return ResponseEntity.created(location).build();
    }
}
```

（9）创建一个 Application 类，该类仅运行我们的应用程序，并创建必要的 bean，这些 bean 会连接到我们的控制器中，具体如下：

```
package com.packtpub.microservices.ch05.message;

import com.packtpub.microservices.ch05.message.clients.SocialGraphClient;
```

```java
import org.springframework.boot.SpringApplication;
import org.springframework.boot.autoconfigure.SpringBootApplication;
import org.springframework.context.annotation.Bean;

@SpringBootApplication
public class Application {
    @Bean
    public MessageRepository messageRepository() {
        return new MessageRepository();
    }

    @Bean
    public SocialGraphClient socialGraphClient() {
        return new SocialGraphClient("http://localhost:4567");
    }

    public static void main(String[] args) {
        SpringApplication.run(Main.class, args);
    }
}
```

上述服务现在可以正常工作，并且满足了我们的主要要求，即如果发送者和接收者不是好友，则无法发送消息，但是它很容易受到我们上述描述的所有问题的影响。例如，如果 social-graph-service 服务遇到问题，则 message-service 服务将取决于 RestTemplate 客户端中的超时设置，这将影响 message-service 服务能够满足的请求数量。此外，如果 social-graph-service 服务不堪重负，并开始返回 503（HTTP 状态码，表示该服务暂时不可用），则 message-service 服务也没有机制允许 social-graph-service 服务恢复。因此，接下来我们将引入 resilience4j 断路器库，并包装对 social-graph-service 服务的调用，步骤如下所示。

（1）打开 build.gradle 并将 resilience4j 断路器库添加到依赖项列表中，具体如下：

```
...
dependencies {
    testCompile group: 'junit', name: 'junit', version: '4.12'
    compile group: 'io.github.resilience4j', name: 'resilience4j-circuitbreaker', version: '0.11.0'
    compile group: 'org.springframework.boot', name: 'spring-boot-starter-web'
}
...
```

（2）修改 SocialGraphClient 以便在调用 social-graph-client 服务时使用 CircuitBreaker。

如果 SocialGraphClient 返回失败，则将返回一个空的 Friendships 实例，这将导致我们的服务以 403 禁止发送代码响应用户请求（默认是关闭的）。我们将在此处使用断路器的默认配置，但是你应该查阅 resilience4j 的说明文档，其中包含有关配置断路器以适合你的特定服务需求的大量信息。来看下列代码：

```
package com.packtpub.microservices.ch05.clients;

import com.packtpub.microservices.ch05.models.Friendships;
import io.github.resilience4j.circuitbreaker.CircuitBreaker;
import io.github.resilience4j.circuitbreaker.CircuitBreakerRegistry;
import io.vavr.CheckedFunction0;
import io.vavr.control.Try;
import org.springframework.web.client.RestTemplate;

import java.util.List;

public class SocialGraphClient {
    private String baseUrl;

    private CircuitBreaker circuitBreaker;

    public SocialGraphClient(String baseUrl) {
        this.baseUrl = baseUrl;
        this.circuitBreaker =
CircuitBreaker.ofDefaults("socialGraphClient");
    }

    public List<String> getFriendships(String username) {

        CheckedFunction0<Friendships> decoratedSupplier =
CircuitBreaker.decorateCheckedSupplier(circuitBreaker, () -> {
            String requestUrl = baseUrl + "/friendships/" + username;
            RestTemplate template = new RestTemplate();
            return template.getForObject(requestUrl,
Friendships.class);
        });

        Try<Friendships> result = Try.of(decoratedSupplier);

        return result.getOrElse(new
```

```
Friendships(username)).getFriendships();
    }
}
```

现在，我们的服务已经将危险的网络调用包装在断路器中，从而防止了social-graph-service 服务的故障级联传导到 message-service 服务。如果 social-graph-service 服务暂时失败，则 message-service 服务最终将很快失败，并使 social-graph-service 服务获得恢复的时间。你可以通过强制模拟 social-graph-service 服务返回错误代码来进行测试，这对于你来说应该是一个有趣的练习。

5.3 使用指数退避算法重试请求

分布式系统中的故障是不可避免的，所以，与其试图完全防止故障，不如设计一种能够自我修复的系统。

5.3.1 理论阐释

在分布式系统中，出现故障并重试时，必须要有一个良好的策略供客户端遵循。例如，服务可能暂时不可用，或出现需要呼叫工程师手动解决的问题。在这两种情况下，客户端都应该能够排队，然后重试请求，以便获得最大的成功机会。

如果发生错误，则不停地重试并不是有效的策略。想象一下，服务刚开始时可能会遇到一个高于常态的故障率，甚至可能使100%的请求失败。在这种情况下，如果客户端不间断地进行重试，那么你将面临一个惊群效应（Thundering-Herd）的问题（就好像银行几位客户取不出钱最后可能会导致挤兑一样），客户端将无限制地重试请求。随着故障时间线的发展，更多的客户端会遇到故障，从而导致更多次的重试。你最终将得到一种如图5-2所示的流量模式，该图与拒绝服务攻击中看到的图类似，并且最终结果都是一样的：由于服务不堪重负而导致级联故障，合法流量也被大量堵塞。你的应用程序将变得无法使用，并且失败的服务将更加难以被隔离和修复。

防止惊群效应的方法是添加一个退避算法，该算法成倍增加重试之间的等待时间，并在发生一定数量的故障后放弃。这种方法被称为上限指数退避（Capped Exponential Backoff）。在重试之间添加呈指数增长的睡眠函数，这是我们策略的前半部分——客户端将减慢重试次数，并随时间分配负载。糟糕的是，客户端的重试仍将是成群结队的，从而导致一段时间内你的服务受到许多并发请求的影响。我们策略的后半部分是通过向睡眠函数添加随机值或抖动（Jitter）来分配重试次数，从而解决此问题。

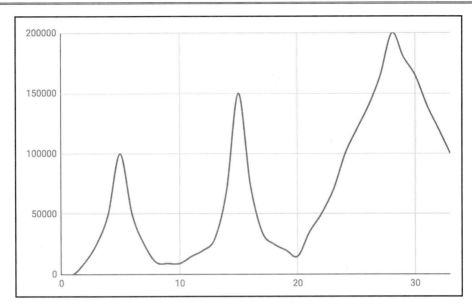

图 5-2

总而言之，我们的重试策略有以下 3 个要求。

- 重试必须使用指数退避算法来隔开。
- 重试必须通过添加抖动来随机化。
- 重试必须在一定的时间后终止。

大多数 HTTP 库都支持满足这些要求的重试策略。在本秘笈中，我们将研究由 Google 编写的 Java 的 HTTP client 库。

5.3.2 实战操作

本秘笈需要执行以下操作。

（1）为了演示如何使用指数退避算法和抖动，我们将在 Ruby 中创建一个示例服务，该服务有一个简单的工作，即返回指示失败的 HTTP 状态。在先前的秘笈中，我们使用了 sinatra Ruby 库执行此操作，因此在这里我们将继续使用它。该服务仅针对每个请求返回 503 HTTP 状态代码，具体如下：

```
require 'sinatra'

get '/' do
    halt 503
end
```

（2）使用 Google HTTP client 库创建一个 HTTP 客户端。首先，使用以下 build.gradle 文件创建一个新的 Gradle Java 项目，该文件将导入必要的库和插件，具体如下：

```gradle
group 'com.packtpub.microservices'
version '1.0-SNAPSHOT'

apply plugin: 'java'
apply plugin: 'application'

mainClassName = 'com.packtpub.microservices.ch05.retryclient.Main'

sourceCompatibility = 1.8

repositories {
    mavenCentral()
}

dependencies {
    compile group: 'com.google.http-client', name: 'google-http-client', version: '1.23.0'
    testCompile group: 'junit', name: 'junit', version: '4.12'
}
```

（3）创建一个名为 com.packtpub.microservices.ch05.retryclient 的新程序包，并创建一个名为 Main 的新类。在 Main 类中，我们将创建一个 HTTP 请求并执行它。如果请求成功，则打印其状态代码，并包含一条成功消息；如果请求失败，则仍然会打印其状态代码，但是包含的消息会指示出现了问题。我们的 HTTP 客户端的第一个版本不会尝试任何重试。这段代码的目的是编写尽可能简单的客户端，而不是炫耀 Google HTTP client 库的功能，但是我们也建议你查阅该项目的说明文档以了解更多信息。来看下面的代码：

```java
package com.packtpub.microservices.ch05.retryclient;

import com.google.api.client.http.*;
import com.google.api.client.http.javanet.NetHttpTransport;
import com.google.api.client.util.ExponentialBackOff;

import java.io.IOException;

public class Main {

    static final HttpTransport transport = new NetHttpTransport();
```

```
    public static void main(String[] args) {
        HttpRequestFactory factory = 
transport.createRequestFactory();
        GenericUrl url = new GenericUrl("http://localhost:4567/");

        try {
            HttpRequest request = factory.buildGetRequest(url);
            HttpResponse response = request.execute();
            System.out.println("Got a successful response: " + 
response.getStatusCode());
        } catch (HttpResponseException e) {
            System.out.println("Got an unsuccessful response: " + 
e.getStatusCode());
        } catch (IOException e) {
            e.printStackTrace();
        }
    }
}
```

（4）如果在集成开发环境（IDE）中运行上述代码（也可以通过命令行 ./gradlew run 运行），则会看到该代码尝试发出单个 HTTP 请求，从 Ruby 服务接收 503，然后放弃。现在让我们用一个可配置的退避算法来进行检测，该退避算法使用一个随机因子来增加抖动，具体如下：

```
package com.packtpub.microservices.ch05.retryclient;

import com.google.api.client.http.*;
import com.google.api.client.http.javanet.NetHttpTransport;
import com.google.api.client.util.ExponentialBackOff;

import java.io.IOException;

public class Main {

    static final HttpTransport transport = new NetHttpTransport();

    public static void main(String[] args) {
        HttpRequestFactory factory = 
transport.createRequestFactory();
        GenericUrl url = new GenericUrl("http://localhost:4567/");
```

```java
    try {
        HttpRequest request = factory.buildGetRequest(url);
        ExponentialBackOff backoff = new ExponentialBackOff.Builder()
                .setInitialIntervalMillis(500)
                .setMaxElapsedTimeMillis(10000)
                .setMaxIntervalMillis(6000)
                .setMultiplier(1.5)
                .setRandomizationFactor(0.5)
                .build();

        request.setUnsuccessfulResponseHandler(
                new HttpBackOffUnsuccessfulResponseHandler(backoff));
        HttpResponse response = request.execute();
        System.out.println("Got a successful response: " + response.getStatusCode());
    } catch (HttpResponseException e) {
        System.out.println("Got an unsuccessful response: " + e.getStatusCode());
    } catch (IOException e) {
        e.printStackTrace();
    }
}
```

如果现在运行程序并查看 Ruby 服务的日志，你将看到代码多次尝试发出请求，并且增加了两次重试之间的睡眠时间，直到大约 10s 后才放弃。在实际环境中，这可以给服务留出足够的时间来恢复，而又不会产生惊群效应导致无法恢复。

5.4 通过缓存提高性能

微服务的设计方式应该是，单个服务往往是对特定数据存储进行读取或写入操作的唯一工具。在此模型中，服务对提供的业务功能所涉及的域模块拥有完全所有权。

5.4.1 理论阐释

具有清晰的边界使开发人员更容易考虑系统中数据的生命周期。我们系统中的某些模块将经常更改，但是许多模块的读取次数远比其写入的次数要多。在这些情况下，我们可以使用缓存来存储那些不经常更改的数据，从而使我们不必在每次请求对象时都向数据库发出请求。数据库查询通常比缓存查询更昂贵，因此理想的方法是尽可能使用缓存。

除了有助于提高性能之外，使用有效的缓存层还可以帮助提高服务的可靠性。我们无法保证数据库的 100%可用性，因此，在数据库发生故障的情况下，服务可以还原为服务缓存的数据。在大多数情况下，相对于完全接收不到数据，用户能够接收一些数据总是更好的（即使数据可能已经过期）。拥有缓存层使你可以配置服务，以将其用作可提供给服务用户的其他可用数据源。

在此秘笈中，我们将创建一个简单的示例服务，该服务将提供有关你的应用程序用户的信息。它有两个端点：第一个端点将接收 POST 请求，并将一个正确格式的用户持久保存到数据库中；第二个端点将按指定的 ID 检索用户信息。ID 存储为 UUID，出于许多原因，它比自动递增的 ID 更可取（在后面的章节中将介绍这些原因）。我们将从基本服务开始，然后添加缓存，以便可以具体了解需要执行哪些步骤。

在此秘笈中，我们将使用 Redis，这是一种流行的开源内存中（In-Memory）数据结构存储，对于存储键值对特别有用。

5.4.2 实战操作

本秘笈需要执行以下操作。

（1）使用以下 build.gradle 文件创建一个名为 caching-user-service 的 Gradle/Java 项目。请注意，我们需要添加 Java 持久化 API（Java Persistence API，JPA）和 Java MySQL client 库的依赖项：

```
group 'com.packtpub.microservices.ch05'
version '1.0-SNAPSHOT'

buildscript {
    repositories {
        mavenCentral()
    }
    dependencies {
        classpath("org.springframework.boot:spring-boot-gradle-plugin:2.0.0.RELEASE")
    }
}

apply plugin: 'java'
apply plugin: 'org.springframework.boot'
```

```
sourceCompatibility = 1.8

repositories {
    mavenCentral()
}

dependencies {
    compile group: 'org.springframework.boot', name: 'spring-boot-
starter-web', version: '2.0.0.RELEASE'
    compile group: 'org.springframework.boot', name: 'spring-boot-
starter-data-jpa', version: '2.0.0.RELEASE'
    compile group: 'mysql', name: 'mysql-connector-java', version:
'6.0.6'
    testCompile group: 'junit', name: 'junit', version: '4.12'
}
```

（2）创建 Main 类。和往常一样，这是应用程序的主要入口，非常简单，具体如下：

```
package com.packtpub.microservices.ch05.userservice;

import org.springframework.boot.SpringApplication;
import org.springframework.boot.autoconfigure.SpringBootApplication;

@SpringBootApplication
public class Main {
    public static void main(String[] args) {
        SpringApplication.run(Main.class, args);
    }
}
```

（3）在 com.packtpub.microservices.ch05.userservice.models 包中创建一个 User 类。这将用作我们的实体表示，并包含将存储在数据库中的字段，它也会存储在 Redis 缓存中，具体如下：

```
package com.packtpub.microservices.ch05.userservice.models;

import com.fasterxml.jackson.annotation.JsonProperty;
import org.hibernate.annotations.GenericGenerator;

import javax.persistence.Column;
import javax.persistence.Entity;
import javax.persistence.GeneratedValue;
```

```java
import javax.persistence.Id;

@Entity
public class User {

    @Id
    @GeneratedValue(generator = "uuid")
    @GenericGenerator(name = "uuid", strategy = "uuid2")
    private String id;

    private String username;

    @JsonProperty("full_name")
    private String fullName;

    private String email;

    public User() {}

    public String getId() {
        return id;
    }

    public void setId(String id) {
        this.id = id;
    }

    public String getUsername() {
        return username;
    }

    public void setUsername(String username) {
        this.username = username;
    }

    public String getFullName() {
        return fullName;
    }

    public void setFullName(String fullName) {
        this.fullName = fullName;
    }
```

```
    public String getEmail() {
        return email;
    }

    public void setEmail(String email) {
        this.email = email;
    }
}
```

(4)要将 User 实体连接到 MySQL 数据库,需要创建一个 UserRepository 接口,该接口扩展了 springframework 数据包定义的 CrudRepository 接口,具体如下:

```
package com.packtpub.microservices.ch05.userservice.db;

import com.packtpub.microservices.ch05.userservice.models.User;
import org.springframework.data.repository.CrudRepository;

public interface UserRepository extends CrudRepository<User, String> {}
```

(5)创建 UserController 类。本示例使用的名称是 RestController,它将某些端点映射到前面讨论的功能,即创建和检索用户记录。接下来的代码你应该似曾相识。值得一提的是,findById 方法将返回 Optional <T>,因此可以使用 map 返回 200 OK HTTP 状态代码以及用户的响应正文,或者使用 orElseGet 返回 404 状态代码,具体如下:

```
package com.packtpub.microservices.ch05.userservice.controllers;

import com.packtpub.microservices.ch05.userservice.db.UserRepository;
import com.packtpub.microservices.ch05.userservice.models.User;
import org.springframework.beans.factory.annotation.Autowired;
import org.springframework.http.HttpStatus;
import org.springframework.http.ResponseEntity;
import org.springframework.web.bind.annotation.*;

import java.util.Optional;

@RestController
public class UserController {

    @Autowired
    private UserRepository userRepository;

    @RequestMapping(path = "/users", method = RequestMethod.POST,
```

```
    produces = "application/json")
    public User create(@RequestBody User user) {
        User savedUser = userRepository.save(user);
        return savedUser;
    }

    @RequestMapping(path = "/users/{id}", method =
RequestMethod.GET, produces = "application/json")
    public ResponseEntity<User> getById(@PathVariable("id") String
id) {
        Optional<User> user = userRepository.findById(id);

        return user.map(u -> new ResponseEntity<>(u,
HttpStatus.OK)).orElseGet(
                () -> new ResponseEntity<>(HttpStatus.NOT_FOUND));
    }
}
```

（6）将以下 application.properties 文件添加到 src/main/resources 目录中。它包含连接到本地 MySQL 实例所需的配置。假定你已安装 MySQL 并使其在本地运行。你还应该创建一个名为 users 的数据库、一个用户名为 userservice 的用户，以及一个密码 password。请注意，我们将 ddl-auto 设置为 create，这对于开发是一种很好的习惯，但它不应用于生产环境：

```
spring.jpa.hibernate.ddl-auto=create
spring.datasource.url=jdbc:mysql://localhost:3306/users?serverTimez
one=UTC&&&useSSL=false
spring.datasource.username=userservice
spring.datasource.password=password
```

（7）现在需要添加一些缓存。再次打开 application.properties 文件，并为在端口 6379（默认）上本地运行的 redis 实例添加一些配置，具体如下：

```
spring.jpa.hibernate.ddl-auto=create
spring.datasource.url=jdbc:mysql://localhost:3306/users?serverTimez
one=UTC&&&useSSL=false
spring.datasource.username=userservice
spring.datasource.password=password
spring.cache.type=redis
spring.redis.host=localhost
spring.redis.port=6379
```

（8）将应用程序配置为使用 MySQL 作为主要数据源，并使用 Redis 作为缓存。现在可以覆盖 CrudRepository<T, ID>接口中的方法，并添加注解以指示其进行缓存。我们希望每次使用 User 对象调用 save 方法时都写入高速缓存中，并在每次使用有效的用户 ID 字符串调用 findById 时从高速缓存中进行读取：

```
package com.packtpub.microservices.ch05.userservice.db;

import com.packtpub.microservices.ch05.userservice.models.User;
import org.springframework.cache.annotation.CachePut;
import org.springframework.cache.annotation.Cacheable;
import org.springframework.data.repository.CrudRepository;
import org.springframework.stereotype.Repository;

import java.util.Optional;

@Repository
public interface UserRepository extends CrudRepository<User, String> {
    @Override
    @Cacheable(value = "users", key = "#id")
    Optional<User> findById(String id);

    @Override
    @CachePut(value = "users", key = "#user.id")
    User save(User user);
}
```

（9）现在可以通过运行服务、创建用户来验证用户记录是否同时位于 MySQL 数据库和 Redis 缓存中，你可以从数据库中删除该用户的记录，然后进行测试，对 users/ID 端点的请求仍将返回用户记录。不过，在完成此服务之前，如果要删除用户，你需要确保缓存不会被修改，因为任何增删用户的端点操作都会使缓存被重写。这项测试任务就留给你作为练习吧。

5.5 通过 CDN 提供更高效的服务

内容交付网络（Content Delivery Network，CDN）也称为内容分发网络，可通过全球分布的代理服务器网络交付内容，从而提高服务的性能和可用性。当用户（通常是使用其移动设备）通过 CDN 向你的 API 发出请求时，他们将根据其地理位置使用多个存在点（Point of Presence，PoP）之一创建网络连接，这样就不必将每个请求都往返到原始数据中心中，而是可以将内容缓存在 CDN 的边缘，从而大大减少了用户的响应时间，并减少了不必要的、昂贵的原始流量。

5.5.1 理论阐释

如果你打算拥有全球用户群，则 CDN 是必需的。如果对应用程序 API 的每个请求都必须执行到单一来源的完整往返，那么你将为与托管应用程序的数据中心物理距离较远的世界上某些地区的用户创造出低于标准的体验。即使你将应用程序托管在多个数据中心中，也无法像 CDN 那样为尽可能多的用户提供尽可能高的性能。

除性能外，CDN 还可以提高应用程序的可用性。正如我们在前面的秘笈中所讨论的那样，系统中许多实体的读取频率远高于它们的写入频率。在这些情况下，你可以将 CDN 配置为在特定时间量内缓存服务的有效负载。这个特定时间量通常由生存时间（Time-To-Live，TTL）指定。缓存来自服务的响应可以减少到你的源数据中心的流量，从而使耗尽资源容量（计算、存储或网络）的情况不容易出现。

此外，如果你的服务开始出现高延迟或者全部或部分失败，则可以将 CDN 配置为提供缓存的响应，而不是继续将流量发送给失败的服务。这使你至少能够在服务中断时向用户提供内容。

某些 CDN 提供商具有允许你自动使资源无效的 API。在这些情况下，你可以检测微服务以使资源无效，就像使用基于 Redis 或基于 Memcached 的缓存一样，这在 5.4 节的秘笈中已经讨论过。

当前有许多不同的 CDN 提供商。其中一些较大的企业包括 Akamai 和 Edgecast。Amazon Web Services 提供了一种被称为 CloudFront 的 CDN 产品，可以将其配置为向 AWS 中的原始服务器或 S3 存储桶中托管的静态资源提供请求服务。CDN 市场上对开发人员更友好的产品之一来自一家名为 Fastly 的公司。Fastly 使用 Varnish 构建，而 Varnish 则是开源的 Web 应用程序加速器。

作为一家提供商，Fastly 允许你上传自己的 Varnish 配置语言（Varnish Configuration Language，VCL）文件，从而有效地允许你基于请求的任何方面（传入的标头、路径段、查询字符串参数等）创建缓存规则。

此外，Fastly 还提供了快速清除 API（Fast Purge API），该 API 可以让开发人员基于 URI 使资源无效。

在本秘笈中，我们将详细介绍创建 CDN 提供商的账号，并开始通过 CDN 提供流量。我们将通过一个假设的服务来做到这一点，该服务可以通过主机名 api.pichat.me 访问公共 Internet。此外，该服务还通过检查传入请求的 Authorization 标头的值以获取有效的

OAuth2 承载令牌来对请求进行身份验证。

5.5.2 实战操作

本秘笈需要执行以下操作。

（1）创建一个 Fastly 账号（Fastly 是一家美国内容交付网络提供商），其注册网址如下：

https://www.fastly.com/signup

（2）Fastly 会要求你创建服务。你需要输入服务的名称、域（api.pichat.me）和运行应用程序的原始服务器的主机名。

（3）使用你的域的 DNS 提供程序，为 api.pichat.me 创建一个 CNAME，将你的域指向 Fastly 的服务器。请阅读说明文档，以了解要使用的主机名。

（4）在设置完成并创建服务之后，对主机名的请求现在将通过 Fastly CDN 进行。请阅读 Fastly 说明文档以了解如何为你的服务自定义 VCL 和其他设置。其网址如下：

https://docs.fastly.com/guides/basic-setup/

5.5.3 优雅地降低用户体验

现在我们已经理解，一定程度的失败是不可避免的。在足够复杂的系统中，某些时间发生一些故障并不稀奇。通过使用本章中介绍的技术，我们可以尝试减少这些可能会影响到客户的故障发生的可能性。但是，无论我们如何努力阻止，某些类型的故障都很可能会在应用程序运行的某个时刻发生并影响客户体验。当然，如果我们能够优雅地降低用户体验，那么面对系统中断之类的突发状况，用户反而可能会抱持同情和理解的态度。

想象一下：你正在使用一个应用程序，该应用程序允许你浏览产品目录并查找售卖该产品的本地商店以及重要信息，如其地址、电话号码和营业时间等。假设提供有关本地商店信息的服务不可用，这显然是以很不理想的方式影响了用户体验，但是应用程序可以通过多种方式处理该故障。

- ❑ 最糟糕的方式（可能会导致最糟糕的用户体验）是让该故障进行级联传导，从产品目录中下架该产品。
- ❑ 好一点的方式是允许用户继续搜索产品，但是当他们去寻找售卖该产品的本地商店时，会通过某种信息框告知他们当前本地商店信息不可用。这也令人沮丧，但是至少他们仍然可以查看到产品信息，如价格、型号和颜色等。

- 更好的方式是识别出该服务未正常运行，然后出现某种信息性提示，告知用户本地商店信息暂时不可用。借助这些信息，我们可以让用户知道当前状态，使他们可以决定是否仍然要继续搜索产品。这样的体验虽然不是最理想的，但是也不会让用户感到很沮丧。

5.6 通过游戏日演习验证容错能力

本章包含的秘笈应可帮助你创建更可靠、更具弹性的微服务架构。每个秘笈都详细说明了一种用于预测和处理某种故障情况的模式或技术。我们构建弹性系统的目的是容忍故障，并且对用户的影响尽可能小。在构建分布式系统时，对故障进行预测和设计是至关重要的，但是如果不验证我们的系统是否能够按照我们期望的方式处理故障，那么我们所做的就仅止步于希望，而希望有可能只是一厢情愿，它绝不是一种有效的策略。

在构建系统时，单元测试和功能测试是我们构建可信工具包的必要组成部分。但是，仅有这些工具是不够的。单元测试和功能测试通过隔离依赖关系来工作，例如良好的单元测试不依赖网络条件，而功能测试则不涉及在生产环境的流量条件下进行测试，而是专注于在理想情况下可以正常工作的各种软件组件。因此，为了对系统的容错能力更有信心，有必要观察其对生产环境中故障的响应。

5.6.1 理论阐释

游戏日演习（Gameday Exercise）是建立系统弹性的另一个有用工具。这些演习的做法是在生产环境中强制执行某些故障方案，以验证我们对容错性的假设是否符合实际情况。John Allspaw 在他的论文 *Fault Injection in Production*（在生产中注入故障）中详细描述了这种做法。如果我们接受"故障无法完全避免"的观念，则人为注入故障并观察系统的应对方式就是一种非常明智的做法。让系统在整个团队密切注视并准备采取行动的情况下出现故障，远比凌晨 3 点系统警报唤醒一名睡眼朦胧的待命工程师要好得多。

计划游戏日演习可提供大量有价值的东西。工程师们应该聚在一起，集体讨论其服务可能遇到的各种故障情况。然后应安排工作以尝试减少或消除这些情况的影响（例如，如果数据库发生故障，则还原为缓存）。每个游戏日演习都应有一个计划文档，其中描述了要测试的系统、各种故障场景（包括将要模拟故障的步骤）、围绕系统应如何应对故障的预期以及对用户的预期影响（如果有的话）。随着游戏日演习的进行，团队应研究每个场景并记录观察结果——重要的是要确保我们期望看到的度量指标已发出，我们希望触发的警报也确实触发了，并且按我们期望的方式处理了故障。在进行观察时，记录下期望与现实之间的任何差异。这些观察应成为有计划的工作，以帮助弥合理想世界与现实世界之间的鸿沟。

本秘笈没有代码可以提供，但是提供了可用于进行游戏日演习的过程和模板。需要说明的是，以下模板不是进行游戏日演习的唯一方法，但它可以作为组织执行此类演习的一个很好的参考。

5.6.2　先决条件

与往常一样，你必须满足一些先决条件，然后才能尝试进行游戏日演习。特别是，你的团队应该习惯于使用必要的度量标准和警报来检测代码，以便在生产环境中提供良好的可观察性。你的团队应具有在事件响应流程中工作的经验并且对该流程有充分的了解，其中包括定期进行回顾，以根据生产事件进行持续改进。

最后，你的组织应该习惯于公开谈论失败和意外的生产事件——不能讳疾忌医或敷衍了事，并且应致力于鼓励持续改进的过程。这些先决条件意味着你的团队应拥有必要的组织支持和心理安全性，以进行此类容错弹性演习。

5.6.3　实战操作

来看以下操作。

（1）游戏日演习的第一步是选择要测试的系统。当你刚开始进行游戏日演习时，明智的做法是选择一个易于理解的系统，该系统以前曾发生过故障，并且万一真的出现较大故障，对用户的影响也有限。

（2）一旦选择了服务，就可以召集负责其开发和运营的团队，并开始集思广益解决不同的故障情况。例如：

如果有数据存储，应考虑由于硬件故障而突然不可用时，会发生什么情况？也许可以手动关闭数据库。

如果数据库以不安全的方式终止，又会发生什么？

假设该服务以某种集群配置运行，因此如果从负载均衡器中删除一个节点，会发生什么情况？

当所有节点都发生故障并从负载平衡池中删除时，会发生什么？

另一个要测试的方面是意外的延迟。在分布式系统中，无法将足够高的延迟与缺乏服务可用性区分开来，因此这里可能会潜藏着很多错误。

团队可以聚在一起讨论所有这些场景（以及其他你可能想到的任何场景），这可能让团队更好地了解系统的好方法。别忘记将你计划测试的所有方案都记录下来。

（3）安排游戏日演习的时间和空间（如果你组织的是远程团队，则需要安排所有人一起进行视频通话）。邀请负责测试服务的团队、客户支持团队的代表以及对演习感兴趣的其他所有利益相关者。

（4）使用模板（可参考 5.6.4 节提供的模板）详细规划演习的进行方式。在计划时间的当天，从被测系统的概述开始。这是一个很好的机会，可确保每个人对系统的工作方式具有一致的看法。然后遍历每种情况，将实际操作分配给团队中的某人。

（5）在演习期间记录观察结果，详细说明系统对故障注入的反应。

（6）如果演习期间观察到的结果与预期不同，则以注记的形式安排后续任务，以便团队纠正差异。

5.6.4　游戏日演习的模板

以下模板可用于计划和执行游戏日演习。

- ❑ 系统名称：

消息服务。

- ❑ 系统总览：

对于被测系统的详细描述（可能包括图表），最好记录下如何将请求路由到系统、与之交互的一些主要系统、其使用的数据存储及其一般配置，以及它所依赖的任何下游服务。

- ❑ 仪表板：

链接到重要的仪表板，以便在游戏日演习进行时观看。

- ❑ 测试方案：
  - ➢ 场景：
    由于节点终止，数据库变得不可用。
  - ➢ 方法：
    使用 AWS CLI 工具（包括实际命令）手动关闭数据库 EC2 节点。
- ❑ 期望：

列出你期望服务如何反应。包括有关指标的预期变化、应触发的警报、系统行为和用户影响的详细信息。

- ❑ 观察：

在实际测试中记录观察结果。

- ❑ 后续行动项目：

为演习后应执行的任何后续工作创建注记。

5.7 引入自动化混沌工程

运行手动的游戏日演习是引入故障注入实践的好方法。在生产环境中人为注入故障有助于建立对系统弹性的信心，并确定改进的机会。游戏日演习可帮助团队在面对许多故障情况时更好地全面了解其系统的行为。随着团队进行更多的演习，它将开始积累用于执行常见任务的工具，例如在网络中引入延迟或让 CPU 使用率产生尖峰。使用工具有助于将单调的任务自动化，提高游戏日演习的效率。有多种开放源代码和商业工具可用于自动化混沌工程（Chaos Engineering），团队可以立即利用这些工具。

5.7.1 理论阐释

比赛日的演习是有计划和调度的。一些组织更进一步引入了连续故障注入，以确保系统平稳地处理常见故障场景。2011 年年初，Netflix 宣布创建了 Simon Army，这是一套旨在将常见故障注入生产环境的工具。Simon Army 中最著名的成员 Chaos Monkey（混沌猴，听起来像是指喜欢调皮捣蛋的孙悟空）会在生产环境中随机关闭节点。Simian Army 工具已开源，可在你自己的组织中使用。除了 Chaos Monkey 之外，Simon Army 中的其他成员如下所示。

- Latency Monkey：在 RESTful 服务的调用中引入人为的延时来模拟服务降级。
- Conformity Monkey：查找不符合最佳实践的实例，并将其关闭。
- Doctor Monkey：查找不健康实例的工具，一旦发现不健康实例，就会将其移出服务组。
- Janitor Monkey：查找不再需要的资源，并将其回收，这能在一定程度上降低云资源的浪费。
- Security Monkey：这是 Conformity Monkey 的一个扩展，可检查系统的安全漏洞。
- 10-18 Monkey：进行本地化及国际化的配置检查。
- Chaos Gorilla：这是 Chaos Monkey 的升级版，可以模拟整个 Amazon Availability Zone 故障。

自动混沌工具可以安排为在游戏日演习中运行，也可以设置为按特定的时间表运行（即星期一至星期五，上午 9 点至下午 5 点，通常是值班工程师在办公室的时间）。

自 2013 年以来，该领域的先驱 PagerDuty 就进行了 Failure Fridays（搞乱星期五）活动。每个星期五，工程师们都聚在一起攻击特定的服务。随着时间的推移，工程师们开始在其聊天机器人中构建命令，以执行常见功能，例如将节点与其他网络流量隔离，甚至添加 Roulette（轮盘赌）命令，该命令将随机选择要重启的主机。

当前已经有人开发出了托管商业服务来帮助自动化混沌工程。Gremlin 就是这样一款托管产品，它可以提供对"攻击"库的访问，通过安装在你环境中节点上的代理执行攻击，以帮助团队运行游戏日演习。Gremlin 提供了一个 API 和一个 Web 界面，该界面使用户能够配置旨在提高资源（CPU、内存、磁盘）使用率的攻击，通过杀死进程或重新启动主机来模拟随机故障以及模拟常见的网络条件，如延迟和网络时间协议（Network Time Protocol，NTP）漂移。拥有 Gremlin 这样的产品可以减少开始进行故障注入所需的前期工作量。

另外还有一个开源工具是 Chaos toolkit，这是一个命令行界面（Command Line Interface，CLI）工具，旨在使设计和运行实验更加容易。在本秘笈中，我们将安装 Chaos toolkit，并使用它对假设的用户服务执行简单的实验。用户服务将与我们在 5.4 节"通过缓存提高性能"中编写的服务相同。

5.7.2 实战操作

本秘笈需要执行以下操作。

（1）Chaos toolkit 是用 Python 编写的，因此可以使用 pip 进行安装。我们需要一个有效的 Python 3 环境。本秘笈将假定你使用 Homebrew 在 macOS X 系统上安装它。首先，安装 pyenv，这是一种支持管理多个 Python 开发环境的实用程序，具体如下：

```
$ brew install pyenv
```

（2）通过执行以下命令行安装 Python 3：

```
$ pyenv install 3.4.2
$ pyenv global 3.4.2
```

（3）对于新安装的 Python 3 环境，请继续执行以下命令行安装 Chaos toolkit：

```
$ pip install -U chaostoolkit
```

（4）Chaos toolkit 使用 JSON 文件描述实验。每个实验都应该有标题、描述信息以及可选的一些用于对实验进行分类的标签。steady-state-hypothesis 部分介绍了正常情况下服务的预期行为。在本示例中，我们假设该服务将在找到用户的情况下返回 200，或者在未找到用户的情况下返回 404，具体如下：

```
{
    "title": "Kill MySQL process",
    "description": "The user service uses a MySQL database to store
```

```json
user information. This experiment will test how the service behaves
when the database is unavailable.",
    "tags": [
        "database", "mysql"
    ],
    "steady-state-hypothesis": {
        "title": "Service responds when MySQL is running",
        "probes": [
            {
                "type": "probe",
                "name": "service-is-running",
                "tolerance": [200, 404],
                "provider": {
                    "type": "http",
                    "url": "http://localhost:8080/users/12345"
                }
            }
        ]
    },
    "method": [
        {
            "name": "kill-mysql-process",
            "type": "action",
            "provider": {
                "type": "process",
                "path": "/usr/local/bin/mysql.server",
                "arguments": ["stop"],
                "timeout": 10
            }
        }
    ]
}
```

（5）执行以下命令运行此实验：

```
$ chaos run
```

（6）如果实验运行成功，则输出应指示当 MySQL 不可用时该服务响应良好。当然，在当前状态下，该实验将使 MySQL 停止运行，这并不是理想的选择。这个问题将作为一项练习，留给读者尝试解决。完成之后，你可以重新运行实验。恭喜！你已经成功运行了你的第一个自动化混沌实验。

第 6 章 安 全 性

本章包含以下操作秘笈。
- 身份验证微服务。
- 确保容器安全。
- 安全配置。
- 安全日志记录。
- 基础架构即代码。

6.1 导 语

与本书中讨论的许多主题一样，微服务架构中的安全性也是需要权衡取舍的。在微服务架构中，单个代码库的职责有限。如果攻击者能够破坏一个正在运行的服务，那么他们将只能执行由该特定微服务控制的操作。但是，微服务架构的分布式性质意味着，攻击者有更多潜在的目标可资利用，它们分布在各个集群运行的服务中。这些集群之间的网络流量（包括边缘服务和内部服务之间的流量）为攻击者提供了许多发现漏洞的机会。

由于微服务架构的分布式性质，因此在配置服务之间的通信方式时必须考虑网络拓扑。单体架构代码库中也存在此问题，因为单体架构代码库的运行实例同样需要通过网络与数据库服务器、缓存、负载平衡器等进行通信。只不过，微服务架构使这些挑战更加明显，因此工程师们必须更早地考虑它们。

安全性是一个宏大话题。本章将讨论在构建、部署和操作微服务时要考虑的许多良好做法，但重要的是要注意，这并不是一个详尽的要考虑事项清单。开发任何系统时都应考虑良好的 API 实践和深入的防御，微服务也不例外。在此我们衷心推荐开放式 Web 应用程序安全项目（Open Web Application Security Project，OWASP）提供的有关 Web 应用程序安全性的资源。其网址如下：

https://www.owasp.org/index.php/Main_Page

6.2 身份验证微服务

在本书第 1 章 "单体架构应用程序分解"中，我们介绍了 Ruby on Rails 代码库，该库为我们虚构的图像消息应用程序 pichat 提供了动力。

6.2.1 理论阐释

Rails 代码库通过检查 Authorization 标头对每个请求进行身份验证。如果存在该标头，则应用程序将尝试使用从环境变量读取的共享机密对其进行解码（详见 6.4 节"安全配置"）。如果 Authorization 标头中提供的令牌有效，则解码后的值将包含有关用户的上下文信息，包括用户 ID。然后，该信息将用于从数据库中检索用户，这样应用程序就可以获得发出请求的用户的上下文信息。如果缺少 Authorization 标头或无法成功解码，则应用程序将引发异常，并向调用方返回 HTTP 401，包括错误消息。为了获得包含在 Authorization 标头中的令牌，客户端应用程序可以使用有效的用户凭据将 POST 请求发送到/auth/login 端点。以下 CURL 命令演示了此流程：

```
$ curl -D - -X POST http://localhost:9292/auth/login -d'email=p@eval.ca&password=foobar123'

HTTP/1.1 200 OK
Content-Type: application/json; charset=utf-8
ETag: W/"3675d2006d59e01f8665f20ffef65fe7"
Cache-Control: max-age=0, private, must-revalidate
X-Request-Id: 6660a102-059f-4afe-b17c-99375db305dd
X-Runtime: 0.150903
Transfer-Encoding: chunked

{"auth_token":"eyJhbGciOiJIUzI1NiJ9.eyJ1c2VyX2lkIjoxLCJleHAiOjE1MzE2ODUxN
jR9.vAToW_mWlOnr-GPzP79EvN62Q2MpsnLIYanz3MTbZ5Q"}
```

现在我们有了令牌，可以将其包含在后续请求的标头中，具体如下：

```
$ curl -X POST -D - -H 'Authorization: 
eyJhbGciOiJIUzI1NiJ9.eyJ1c2VyX2lkIjoxLCJleHAiOjE1MzE2ODUxNjR9.vAToW_mWlOn
r-GPzP79EvN62Q2MpsnLIYanz3MTbZ5Q' http://localhost:9292/messages -
d'body=Hello&user_id=1'

HTTP/1.1 201 Created
```

```
Content-Type: application/json; charset=utf-8
ETag: W/"211cdab551e63ca48de48217357f1cf7"
Cache-Control: max-age=0, private, must-revalidate
X-Request-Id: 1525333c-dada-40ff-8c25-a0e7d151433c
X-Runtime: 0.019609
Transfer-Encoding: chunked

{"id":1,"body":"Hello","user_id":1,"created_at":"2018-07-14T20:08:19.369Z
", "updated_at":"2018-07-14T20:08:19.369Z","from_user_id":1}
```

由于 pichat-api 是单体架构代码库,因此它在支持此流程方面扮演着许多不同的角色。它可以充当授权服务、身份验证网关、用户存储和授权客户端。这正是我们在微服务架构中要避免的责任耦合。

幸运的是,开发人员可以很轻松地将这些职责划分为单独的代码库,同时保持流程相同。我们可以使用共享机密将信息编码为 JSON Web 令牌(JSON Web Token,JWT),使得单个微服务可以安全地验证请求,而不必为每个请求向中心身份验证服务发出请求。获取身份验证令牌可以是中心服务的职责,但是可以使用 API 网关或服务于前端的后端(BFF)使这一事实对客户端透明。图 6-1 演示了如何划分这些职责。

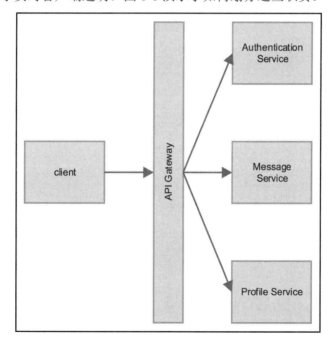

图 6-1

原文	译文	原文	译文
client	客户端	Message Service	消息服务
API Gateway	API 网关	Profile Service	配置文件访问
Authentication Service	身份验证服务		

我们将创建一个身份验证服务,该服务处理用户注册并交换 JWT 的凭据。然后,我们将使用第 2 章"边缘服务"中介绍的 Zuul 开源项目创建一个简单的 API 网关。

6.2.2 实战操作

本秘笈需要执行以下操作。

(1)创建身份验证服务。使用以下 build.gradle 文件创建一个新的 Java 项目:

```
group 'com.packtpub.microservices'
version '1.0-SNAPSHOT'

buildscript {
    repositories {
        mavenCentral()
    }
    dependencies {
        classpath group: 'org.springframework.boot', name: 'spring-boot-gradle-plugin', version: '1.5.9.RELEASE'
    }
}

apply plugin: 'java'
apply plugin: 'org.springframework.boot'
apply plugin: 'io.spring.dependency-management'

sourceCompatibility = 1.8

repositories {
    mavenCentral()
}

dependencies {
    compile group: 'org.springframework.boot', name: 'spring-boot-starter-web'
    compile group: 'org.springframework.security', name: 'spring-security-core'
```

```
    compile group: 'org.springframework.security', name: 'spring-
security-config'
    compile group: 'org.springframework.boot', name: 'spring-boot-
starter-data-jpa'
    compile group: 'io.jsonwebtoken', name: 'jjwt', version:
'0.9.1'
    compile group: 'mysql', name: 'mysql-connector-java'
    testCompile group: 'junit', name: 'junit', version: '4.12'
}
```

我们将用户凭据存储在 MySQL 数据库中，因此可以将 mysql-connector-java 声明为依赖项。我们还将使用一个名为 jjwt 的开源 JWT 库。

提示：

存储用户凭据是一个重要的主题。用户密码绝不能以纯文本格式存储，并且已证明许多哈希算法（如 MD5 和 SHA1）容易受到各种暴力攻击。在本示例中，我们将使用 bcrypt。在实际使用中，我们将考虑多个哈希步骤，例如先使用 SHA512 进行哈希，然后通过 bcrypt 运行。我们还将考虑添加每个用户的盐值（Salt）。开放 Web 应用程序安全项目在存储密码方面有很多不错的建议，其网址如下：

https://www.owasp.org/index.php/Password_Storage_Cheat_Sheet

（2）创建一个名为 Application 的新类。它将包含我们的主要方法以及 PasswordEncoder，具体如下：

```
package com.packtpub.microservices.ch06.auth;

import org.springframework.boot.SpringApplication;
import org.springframework.boot.autoconfigure.SpringBootApplication;
import org.springframework.context.annotation.Bean;
import org.springframework.security.crypto.bcrypt.BCryptPasswordEncoder;
import org.springframework.security.crypto.password.PasswordEncoder;

@SpringBootApplication
public class Application {

    @Bean
    public PasswordEncoder passwordEncoder() {
        return new BCryptPasswordEncoder();
    }
```

```
    public static void main(String[] args) {
        SpringApplication.run(Application.class, args);
    }
}
```

（3）我们将用户凭证建模为带有 email 和 password 字段的简单 POJO。据此，创建一个名为 com.packtpub.microservices.ch06.auth.models 的新程序包和一个名为 UserCredential 的新类，具体如下：

```
package com.packtpub.microservices.ch06.auth.models;

import org.hibernate.annotations.GenericGenerator;

import javax.persistence.*;

@Entity
public class UserCredential {
    @Id
    @GeneratedValue(generator = "uuid")
    @GenericGenerator(name = "uuid", strategy = "uuid2")
    private String id;

    @Column(unique=true)
    private String email;

    private String password;

    public UserCredential(String email) {
        this.email = email;
    }

    public String getId() {
        return id;
    }

    public void setId(String id) {
        this.id = id;
    }

    public String getEmail() {
        return email;
    }
```

```
    public void setEmail(String email) {
        this.email = email;
    }

    public String getPassword() {
        return password;
    }

    public void setPassword(String password) {
        this.password = password;
    }
}
```

(4)创建一个模块来表示对成功登录和注册请求的响应。成功的响应将包含一个 JSON 文档,该文档中包括 JWT。据此,创建一个名为 AuthenticationToken 的新类,具体如下:

```
package com.packtpub.microservices.ch06.auth.models;

import com.fasterxml.jackson.annotation.JsonProperty;

public class AuthenticationToken {

    @JsonProperty("auth_token")
    private String authToken;

    public AuthenticationToken() {}

    public AuthenticationToken(String authToken) {
        this.authToken = authToken;
    }

    public String getAuthToken() {
        return this.authToken;
    }

    public void setAuthToken(String authToken) {
        this.authToken = authToken;
    }
}
```

(5)我们将使用 Java Persistence API 访问 UserCredential 类。为此,必须首先创建

CrudRepository。这里,创建一个名为 com.packtpub.microservices.ch06.auth.data 的新程序包和一个名为 UserCredentialRepository 的新类。除了继承自 CrudRepository 之外,我们还将定义一个用于通过电子邮件检索 UserCredential 实例的方法,具体如下:

```
package com.packtpub.microservices.ch06.auth.data;

import com.packtpub.microservices.ch06.auth.models.UserCredential;
import org.springframework.data.repository.CrudRepository;

public interface UserCredentialRepository extends
CrudRepository<UserCredential, String> {
    UserCredential findByEmail(String email);
}
```

(6)当用户尝试使用无效的凭据注册或登录时,我们希望返回 HTTP 401 状态代码以及一条消息,指出他们提供了无效的凭据。为了做到这一点,我们将创建一个异常,该异常将在我们的控制器方法中抛出,具体如下:

```
package com.packtpub.microservices.ch06.auth.exceptions;

import org.springframework.http.HttpStatus;
import org.springframework.web.bind.annotation.ResponseStatus;

@ResponseStatus(HttpStatus.UNAUTHORIZED)
public class InvalidCredentialsException extends Exception {
    public InvalidCredentialsException(String message) {
super(message);   }
}
```

(7)创建控制器,登录和注册端点将由单个控制器提供服务。其中,注册方法将仅验证输入并创建一个新的 UserCredential 实例,同时使用我们之前创建的 CrudRepository 包将其持久化,然后,它将使用新注册用户的用户 ID 作为主题对 JWT 进行编码;登录方法将验证提供的凭据,并向 JWT 提供用户 ID 作为其主题;控制器将需要访问在主类中定义的 UserCredentialRepository 和 PasswordEncoder;据此,创建一个名为 com.packtpub.microservices.ch06.auth.controllers 的新程序包和一个名为 UserCredentialController 的新类,具体如下:

```
package com.packtpub.microservices.ch06.auth.controllers;

import com.packtpub.microservices.ch06.auth.data.UserCredentialRepository;
import com.packtpub.microservices.ch06.auth.exceptions.
```

```java
InvalidCredentialsException;
import com.packtpub.microservices.ch06.auth.models.AuthenticationToken;
import com.packtpub.microservices.ch06.auth.models.UserCredential;
import io.jsonwebtoken.JwtBuilder;
import io.jsonwebtoken.Jwts;
import io.jsonwebtoken.SignatureAlgorithm;
import org.springframework.beans.factory.annotation.Autowired;
import org.springframework.beans.factory.annotation.Value;
import org.springframework.security.crypto.password.PasswordEncoder;
import org.springframework.web.bind.annotation.*;

import javax.crypto.spec.SecretKeySpec;
import javax.xml.bind.DatatypeConverter;
import java.security.Key;

@RestController
public class UserCredentialController {

    @Autowired
    private UserCredentialRepository userCredentialRepository;

    @Autowired
    private PasswordEncoder passwordEncoder;

    @Value("${secretKey}")
    private String keyString;

    private String encodeJwt(String userId) {
        System.out.println("SIGNING KEY: " + keyString);
        Key key = new SecretKeySpec(
                DatatypeConverter.parseBase64Binary(keyString),
                SignatureAlgorithm.HS256.getJcaName());

        JwtBuilder builder = Jwts.builder().setId(userId)
                .setSubject(userId)
                .setIssuer("authentication-service")
                .signWith(SignatureAlgorithm.HS256, key);

        return builder.compact();
    }

    @RequestMapping(path = "/register", method =
```

```
RequestMethod.POST, produces = "application/json")
    public AuthenticationToken register(@RequestParam String email,
@RequestParam String password, @RequestParam String
passwordConfirmation) throws InvalidCredentialsException {
        if (!password.equals(passwordConfirmation)) {
            throw new InvalidCredentialsException("Password and
confirmation do not match");
        }

        UserCredential cred = new UserCredential(email);
        cred.setPassword(passwordEncoder.encode(password));
        userCredentialRepository.save(cred);

        String jws = encodeJwt(cred.getId());

        return new AuthenticationToken(jws);
    }

    @RequestMapping(path = "/login", method = RequestMethod.POST,
produces = "application/json")
    public AuthenticationToken login(@RequestParam String email,
@RequestParam String password) throws InvalidCredentialsException {
        UserCredential user =
userCredentialRepository.findByEmail(email);

        if (user == null || !passwordEncoder.matches(password,
user.getPassword())) {
            throw new InvalidCredentialsException("Username or
password invalid");
        }

        String jws = encodeJwt(user.getId());
        return new AuthenticationToken(jws);
    }
}
```

（8）因为我们正在连接到本地数据库，并且因为在签名 JWT 时使用了共享密钥，所以需要创建一个很小的属性文件。据此，在 src/main/resources 目录中创建一个名为 application.yml 的文件，具体如下：

```
server:
  port: 8081
```

```yaml
spring:
    jpa.hibernate.ddl-auto: create
    datasource.url: jdbc:mysql://localhost:3306/user_credentials
    datasource.username: root
    datasource.password:

secretKey: supers3cr3t
```

既然我们已经有了一个可以正常运行的身份验证服务，那么下一步就是使用开源网关服务 Zuul 创建一个简单的 API 网关。除了将请求路由到下游服务之外，API 网关还将使用身份验证过滤器来验证在请求的标头中是否传递了有效的 JWT。

（9）使用以下 build.gradle 文件创建一个新的 Java 项目：

```gradle
group 'com.packtpub.microservices'
version '1.0-SNAPSHOT'

buildscript {
    repositories {
        mavenCentral()
    }
    dependencies {
        classpath group: 'org.springframework.boot', name: 'spring-boot-gradle-plugin', version: '1.5.9.RELEASE'
    }
}

apply plugin: 'java'
apply plugin: 'org.springframework.boot'
apply plugin: 'io.spring.dependency-management'

sourceCompatibility = 1.8
targetCompatibility = 1.8

repositories {
    mavenCentral()
}

dependencyManagement {
    imports {
        mavenBom 'org.springframework.cloud:spring-cloud-netflix:1.4.4.RELEASE'
    }
```

```
}
dependencies {
    compile group: 'org.springframework.boot', name: 'spring-boot-
starter-web'
    compile group: 'org.springframework.cloud', name: 'spring-
cloud-starter-zuul'
    compile group: 'org.springframework.security', name: 'spring-
security-core'
    compile group: 'org.springframework.security', name: 'spring-
security-config'
    compile group: 'org.springframework.security', name: 'spring-
security-web'
    compile group: 'io.jsonwebtoken', name: 'jjwt', version:
'0.9.1'
    testCompile group: 'junit', name: 'junit', version: '4.12'
}
```

请注意，我们使用的 JWT 库与身份验证服务相同。

（10）创建一个名为 com.packtpub.microservices.ch06.gateway 的新程序包和一个名为 Application 的新类，具体如下：

```
package com.packtpub.microservices.ch06.gateway;

import org.springframework.boot.SpringApplication;
import org.springframework.boot.autoconfigure.SpringBootApplication;
import org.springframework.cloud.netflix.zuul.EnableZuulProxy;

@EnableZuulProxy
@SpringBootApplication
public class Application {
    public static void main(String[] args) {
        SpringApplication.run(Application.class, args);
    }
}
```

（11）我们将通过创建 OncePerRequestFilter 的子类来创建身份验证过滤器，该子类旨在为每个请求分派提供一次执行。过滤器将从 Authorization 标头中解析 JWT，并尝试使用共享机密对其进行解码。如果可以验证和解码 JWT，则可以确定它是由有权访问共享机密的发行者（Issuer）编码的。我们将其视为我们的信任边界；有权访问共享机密的任何人都是可以信任的，因此我们可以相信 JWT 的主题是已认证用户的 ID。据此，创建一个名为 AuthenticationFilter 的新类，具体如下：

```java
package com.packtpub.microservices.ch06.gateway;

import io.jsonwebtoken.Claims;
import io.jsonwebtoken.Jwts;
import org.springframework.security.authentication.UsernamePasswordAuthenticationToken;
import org.springframework.security.core.Authentication;
import org.springframework.security.core.authority.SimpleGrantedAuthority;
import org.springframework.security.core.context.SecurityContextHolder;
import org.springframework.web.filter.OncePerRequestFilter;
import javax.servlet.FilterChain;
import javax.servlet.ServletException;
import javax.servlet.http.HttpServletRequest;
import javax.servlet.http.HttpServletResponse;
import javax.xml.bind.DatatypeConverter;
import java.io.IOException;
import java.util.ArrayList;
import java.util.Optional;

public class AuthenticationFilter extends OncePerRequestFilter {

    private String signingSecret;

    AuthenticationFilter(String signingSecret) {
        this.signingSecret = signingSecret;
    }

    @Override
    protected void doFilterInternal(HttpServletRequest request, HttpServletResponse response, FilterChain filterChain) throws ServletException, IOException {
        Optional<String> token = Optional.ofNullable(request.getHeader("Authorization"));
        Optional<Authentication> auth = token.filter(t -> t.startsWith("Bearer")).flatMap(this::authentication);
        auth.ifPresent(a -> SecurityContextHolder.getContext().setAuthentication(a));
        filterChain.doFilter(request, response);
    }
```

```java
    private Optional<Authentication> authentication(String t) {
        System.out.println(signingSecret);
        String actualToken = t.substring("Bearer ".length());
        try {
            Claims claims = Jwts.parser()
.setSigningKey(DatatypeConverter.parseBase64Binary(signingSecret))
                .parseClaimsJws(actualToken).getBody();
            Optional<String> userId =
Optional.ofNullable(claims.getSubject()).map(Object::toString);
            return userId.map(u -> new
UsernamePasswordAuthenticationToken(u, null, new
ArrayList<SimpleGrantedAuthority>()));
        } catch (Exception e) {
            return Optional.empty();
        }
    }
}
```

（12）将 Authentication 类与 API 网关项目的安全性配置连接在一起。针对于此，创建一个名为 SecurityConfig 的新类，具体如下：

```java
package com.packtpub.microservices.ch06.gateway;

import org.springframework.beans.factory.annotation.Value;
import org.springframework.security.config.annotation.web.builders.HttpSecurity;
import org.springframework.security.config.annotation.web.configuration.EnableWebSecurity;
import org.springframework.security.config.annotation.web.configuration.WebSecurityConfigurerAdapter;
import org.springframework.security.config.http.SessionCreationPolicy;
import org.springframework.security.web.authentication.UsernamePasswordAuthenticationFilter;

import javax.servlet.http.HttpServletResponse;

@EnableWebSecurity
public class SecurityConfig extends WebSecurityConfigurerAdapter {

    @Value("${jwt.secret}")
    private String signingSecret;
```

```
    @Override
    protected void configure(HttpSecurity security) throws
Exception {
        security
            .csrf().disable()
            .logout().disable()
            .formLogin().disable()
.sessionManagement().sessionCreationPolicy(SessionCreationPolicy.ST ATELESS)
            .and()
                .anonymous()
            .and()
                .exceptionHandling().authenticationEntryPoint(
                    (req, rsp, e) ->
rsp.sendError(HttpServletResponse.SC_UNAUTHORIZED))
            .and()
            .addFilterAfter(new
AuthenticationFilter(signingSecret),
                    UsernamePasswordAuthenticationFilter.class)
            .authorizeRequests()
            .antMatchers("/auth/**").permitAll()
            .antMatchers("/messages/**").authenticated()
            .antMatchers("/users/**").authenticated();
    }
}
```

从上述代码中可以看到，我们允许对身份验证服务的任何请求（以/auth/...为前缀的请求），同时要求对用户或消息服务的请求进行身份验证。

（13）我们需要一个配置文件来存储共享机密以及 Zuul 服务器的路由信息。据此，在 src/main/resources 目录中创建一个名为 application.yml 的文件，具体如下：

```
server:
    port: 8080

jwt:
    secret: supers3cr3t

zuul:
    routes:
        authentication-service:
            path: /auth/**
            url: http://127.0.0.1:8081
        message-service:
```

```
            path: /messages/**
            url: http://127.0.0.1:8082
        user-service:
            path: /users/**
            url: http://127.0.0.1:8083
```

（14）现在已经有了一个能够正常工作的身份验证服务和一个能够验证 JWT 的 API 网关，我们可以通过使用前述配置文件中定义的端口运行 API 网关、身份验证服务和消息服务来测试身份验证方案。现在，以下 CURL 请求表明可以为 JWT 交换有效的凭据，并且可以使用 JWT 访问受保护的资源。我们还可以证明，如果没有有效的 JWT，则对受保护资源的请求将被拒绝。

💡 提示：

请注意，在此示例中，message-service 服务仍然不对请求进行任何授权。理论上，任何提出身份验证请求的人都可以访问其他人的消息。message-service 服务应进行修改，以检查来自 JWT 主题的用户 ID，并且仅允许访问属于该用户的消息。

（15）可以使用 curl 来测试注册新用户账户，具体如下：

```
$ curl -X POST -D - http://localhost:8080/auth/register -
d'email=p@eval.ca&password=foobar123&passwordConfirmation=foobar123'

HTTP/1.1 200
X-Content-Type-Options: nosniff
X-XSS-Protection: 1; mode=block
Cache-Control: no-cache, no-store, max-age=0, must-revalidate
Pragma: no-cache
Expires: 0
X-Frame-Options: DENY
X-Application-Context: application:8080
Date: Mon, 16 Jul 2018 03:27:17 GMT
Content-Type: application/json;charset=UTF-8
Transfer-Encoding: chunked

{"auth_token":"eyJhbGciOiJIUzI1NiJ9.eyJqdGkiOiJmYWQzMGZiMi03MzhmLTR
iM2QtYTIyZC0zZGNmN2NmNGQ1NGIiLCJzdWIiOiJmYWQzMGZiMi03MzhmLTRiM2QtYT
IyZC0zZGNmN2NmNGQ1NGIiLCJpc3MiOiJhdXRoZW50aWNhdGlvbi1zZXJ2aWNlIn0.T
zOKItjBU-AtRMqIB_D1n-qv6IO_zCBIK8ksGzsTC90"}
```

（16）现在我们已经有了一个 JWT，可以将其包含在 message-service 服务请求的标头中，以测试 API 网关是否能够验证和解码令牌，具体如下：

```
$ curl -D - -H "Authorization: Bearer
eyJhbGciOiJIUzI1NiJ9.eyJqdGkiOiI3YmU4N2U3Mi03ZjhhLTQ3ZjktODk3NS1mYz
M5ZTE0NjNmODAiLCJzdWIiOiI3YmU4N2U3Mi03ZjhhLTQ3ZjktODk3NS1mYzM5ZTE0N
jNmODAiLCJpc3MiOiJhdXRoZW50aWNhdGlvbi1zZXJ2aWNlIn0.fpFbHhdSEVKk95m5
Q7iNjkKyM-eHkCGGKchTTKgbGWw" http://localhost:8080/messages/123

HTTP/1.1 404
X-Content-Type-Options: nosniff
X-XSS-Protection: 1; mode=block
Cache-Control: no-cache, no-store, max-age=0, must-revalidate
Pragma: no-cache
Expires: 0
X-Frame-Options: DENY
X-Application-Context: application:8080
Date: Mon, 16 Jul 2018 04:05:40 GMT
Content-Type: application/json;charset=UTF-8
Transfer-Encoding: chunked

{"timestamp":1532318740403,"status":404,"error":"Not
Found","exception":"com.packtpub.microservices.ch06.message.
exceptions.MessageNotFoundException","message":"Message 123
could not be found","path":"/123"}
```

从上述代码中可以看到，我们从 message-service 服务中获得 404 状态代码，这一事实表明，请求到达该服务。如果修改请求标头中的 JWT，则应该得到 401，具体如下：

```
$ curl -D - -H "Authorization: Bearer not-the-right-jwt"
http://localhost:8080/messages/123

HTTP/1.1 401
X-Content-Type-Options: nosniff
X-XSS-Protection: 1; mode=block
Cache-Control: no-cache, no-store, max-age=0, must-revalidate
Pragma: no-cache
Expires: 0
X-Frame-Options: DENY
Content-Type: application/json;charset=UTF-8
Transfer-Encoding: chunked
Date: Mon, 23 Jul 2018 04:06:47 GMT

{"timestamp":1532318807874,"status":401,"error":"Unauthorized",
"message":"No message available","path":"/messages/123"}
```

6.3 确保容器安全

容器的出现为管理微服务架构的组织解决了许多问题。容器允许将服务捆绑为一个独立的单元,并且软件及其依赖项可以构建为单个工件,然后打包提交到任何要运行或调度的环境中。容器不再依赖复杂的配置管理解决方案来管理生产系统的细微变化,而是支持不变的基础架构的思想。一旦构建了基础架构,就不必升级或维护它。相反,你只需要构建新的基础架构,然后丢弃旧的基础架构即可。

6.3.1 理论阐释

容器还允许组织优化其对存储和计算资源的使用。

因为可以将软件构建为容器,所以可以在单个虚拟机或硬件上运行多个应用程序,每个虚拟机都不知道其他虚拟机的存在。尽管多租户(Multi-Tenancy)技术具有许多优点,但在同一虚拟机上运行多个服务会引入恶意用户可能利用的新攻击情形。如果攻击者能够利用一项服务中的漏洞,则他们可能利用该漏洞来攻击同一虚拟机上运行的服务。在这种设置中,默认情况下,集群被视为安全边界。如果你有权访问集群,则必须被信任。

根据组织的需求,将集群作为安全边界可能还不够,我们可能还希望在容器之间实现更高的安全性和隔离。seccomp 安全性功能已在 Linux 内核 Version 2.6.12 中引入,它支持限制可以从进程进行的系统调用。使用 seccomp 策略运行容器化的应用程序实际上会将服务和容器中运行的其他进程沙箱化。在本秘笈中,我们将向你展示如何检查是否已在 Linux 内核中配置 seccomp,并演示如何使用自定义 seccomp 策略运行容器。

6.3.2 实战操作

本秘笈需要执行以下操作。

(1)为了对 Docker 容器使用 seccomp 策略,必须在配置了 seccomp 支持的 Linux 内核的主机上运行该容器。要检查这一点,可以在内核配置文件中搜索 CONFIG_SECCOMP,命令如下:

```
$ grep CONFIG_SECCOMP= /boot/config-$(uname -r)
CONFIG_SECCOMP=y
```

(2)现在,我们已经确认在 Linux 内核中已启用 seccomp,接下来可以看看 Docker

打包的默认配置文件,其网址如下:

https://github.com/moby/moby/blob/master/profiles/seccomp/default.json

上述默认策略足以满足大多数需求,并且具有相当的限制性。如果已启用 seccomp 支持,则容器将使用此策略运行。

(3)为了进一步验证 seccomp 已配置并且 Docker 能够支持它,我们将创建一个简单的自定义策略,然后在一个容器中运行一个命令以证明该策略正在执行。据此,创建一个名为 policy.json 的文件,具体如下:

```
{
    "defaultAction": "SCMP_ACT_ALLOW",
    "syscalls": [
        {
            "name": "chown",
            "action": "SCMP_ACT_ERRNO"
        }
    ]
}
```

(4)现在,运行一个执行外壳的容器并尝试创建文件,然后更改所有权。错误消息表明该容器受 seccomp 策略限制,具体如下:

```
$ docker run --rm -it --security-opt seccomp:policy.json busybox /bin/sh
/ # touch foo
/ # chown root foo
chown: foo: Operation not permitted
```

6.4 安全配置

服务通常需要某种形式的配置。服务配置将存储所有信息,这些信息可能会根据部署服务的环境而变化。例如,当在开发人员的工作站上以本地方式运行服务时,该服务可能应连接到同样在本地运行的数据库。但是,在生产环境中,服务应连接到生产数据库。

6.4.1 理论阐释

一般来说,配置中存储的公共数据包括数据存储的位置和凭据、访问令牌或第三方服务和操作信息的其他凭据，例如在初始化连接池或配置网络超时时将度量发送到何处或使用哪些值连接。

将配置与代码分开存储很重要。进行配置更改时，不必将更改提交给源代码存储库，也不必创建新的程序包并运行单独的部署。理想情况下，应该有一种简单的方法来更改配置，而无须部署新版本的服务。从安全性角度来看，将配置存储在代码中（例如，在源代码文件中对密码进行硬编码）也是一种不好的做法，因为有权访问源代码的任何人都可以访问配置，对于机密信息来说，很少有这样处理的。比较好的做法是尽可能频繁地滚动密钥和凭据，以使得即便机密信息被泄露或很容易被泄露，它也不会长期有效。硬编码的机密信息使这种做法变得很困难，所以在实践中基本上不会这样做。

常见的最佳做法是将配置存储在环境变量中。这是将配置值公开给进程的一种好方法，该方式可以根据服务运行的环境轻松更改。环境变量适合非秘密的配置值，例如主机名、超时和日志级别。另外，环境变量一般不足以存储机密信息。

将机密信息存储为环境变量固然让任何进程都可以访问其值（这些进程需要在与服务相同的容器或进程空间中运行），但这也使得它们很容易被拦截。有多种解决方案可将机密信息与应用程序其余配置分开存储。部署在 Kubernetes 集群上的应用程序可以使用一种称为 secret 的特殊对象类型，它就是为此目的设计的。

Kubernetes 机密信息在节点之间传输时会使用主节点上的私钥进行加密，但是，该机密信息在静止时以明文形式存储。理想情况下，机密信息应存储为加密值，并且只能由明确允许这样做的进程解密。

Vault 是 HashiCorp 积极维护的一个开源项目。其目的是提供一种易于使用的系统，用于安全地存储和访问机密信息。除存储机密信息外，Vault 还提供访问日志审核、细粒度的访问控制和轻松滚动。在此秘笈中，我们将创建一个新服务，称为 attachment-service 服务，该服务负责处理消息的图像和视频附件。在上传照片和视频文件时，attachment-service 服务将使用 Vault 获取用于访问 S3 存储桶的有效 AWS 凭证。此外，该服务还将使用 Vault 获取其中将存储附件元数据的 MySQL 数据库的数据库凭据。非敏感配置（例如，数据库名称或将照片和视频上传到的 S3 存储桶的名称）将作为环境变量提供给服务。

6.4.2 实战操作

为了演示如何使用 Vault 安全地存储敏感配置数据，首先将创建一个 attachment-service 服务，该服务使用环境变量存储敏感信息；然后将集成 Vault，以便从安全存储中读取相同的配置。

来看以下操作。

（1）使用以下 build.gradle 文件创建一个名为 attachment-service 服务的新 Java 项目：

```
group 'com.packtpub.microservices'
version '1.0-SNAPSHOT'

buildscript {
    repositories {
        mavenCentral()
    }
    dependencies {
        classpath group: 'org.springframework.boot', name: 'spring-boot-gradle-plugin', version: '1.5.9.RELEASE'
    }
}

apply plugin: 'java'
apply plugin: 'org.springframework.boot'
apply plugin: 'io.spring.dependency-management'

sourceCompatibility = 1.8

repositories {
    mavenCentral()
}

dependencies {
    compile group: 'org.springframework.boot', name: 'spring-boot-starter-web'
    compile group: 'org.springframework.boot', name: 'spring-boot-starter-data-jpa', version: '1.5.9.RELEASE'
    compile group: 'mysql', name: 'mysql-connector-java'
    compile group: 'com.amazonaws', name: 'aws-java-sdk-s3', version: '1.11.375'
    testCompile group: 'junit', name: 'junit', version: '4.12'
}
```

（2）创建一个名为 com.packtpub.microservices.ch06.attachment 的新程序包，并创建一个名为 Application 的新类，它将作为我们服务的入口点。除了运行 Spring Boot 应用程序之外，该类还将公开一个 Bean，即 Amazon S3 客户端。请注意，我们正在使用 EnvironmentVariableCredentialsProvider 类，当前该类从一组环境变量中读取凭据，这不是我们在生产环境中要执行的操作。本步骤对应的代码如下：

```
package com.packtpub.microservices.ch06.attachment;

import com.amazonaws.auth.EnvironmentVariableCredentialsProvider;
import com.amazonaws.regions.Regions;
import com.amazonaws.services.s3.AmazonS3;
import com.amazonaws.services.s3.AmazonS3ClientBuilder;

import org.springframework.boot.SpringApplication;
import org.springframework.boot.autoconfigure.SpringBootApplication;
import org.springframework.context.annotation.Bean;

@SpringBootApplication
public class Application {

    @Bean
    public AmazonS3 getS3Client() {
        AmazonS3ClientBuilder client = AmazonS3ClientBuilder.standard();
        return client.withCredentials(
                new EnvironmentVariableCredentialsProvider()).withRegion(Regions.US_WEST_2).build();
    }

    public static void main(String[] args) {
        SpringApplication.run(Application.class, args);
    }
}
```

（3）创建一个名为 com.packtpub.microservices.ch06.attachment.models 的新程序包和一个名为 Attachment 的新类。这将表示我们存储在关系数据库中的附件。本步骤对应的代码如下：

```
package com.packtpub.microservices.ch06.attachment.models;

import org.hibernate.annotations.GenericGenerator;

import javax.persistence.Column;
import javax.persistence.Entity;
import javax.persistence.GeneratedValue;
import javax.persistence.Id;

@Entity
public class Attachment {
```

```java
    @Id
    @GeneratedValue(generator = "uuid")
    @GenericGenerator(name = "uuid", strategy = "uuid2")
    private String id;

    @Column(unique = true)
    private String messageId;
    private String url;
    private String fileName;
    private Integer mediaType;

    public Attachment(String messageId, String url, String fileName, Integer mediaType) {
        this.messageId = messageId;
        this.url = url;
        this.fileName = fileName;
        this.mediaType = mediaType;
    }

    public String getId() {
        return id;
    }

    public void setId(String id) {
        this.id = id;
    }

    public String getMessageId() {
        return messageId;
    }

    public void setMessageId(String messageId) {
        this.messageId = messageId;
    }

    public String getUrl() {
        return url;
    }

    public void setUrl(String url) {
        this.url = url;
```

```
    }

    public String getFileName() {
        return fileName;
    }

    public void setFileName(String fileName) {
        this.fileName = fileName;
    }

    public Integer getMediaType() {
        return mediaType;
    }

    public void setMediaType(Integer mediaType) {
        this.mediaType = mediaType;
    }
}
```

（4）为了对先前定义的 Attachment 类执行基本操作，我们将创建一个名为 com.packtpub.microservices.ch06.attachment.data 的新程序包和一个名为 AttachmentRepository 的接口，该接口扩展了 CrudRepository。我们还将定义一个自定义方法签名，该签名允许调用方查找与特定消息相关的所有附件，具体如下：

```
package com.packtpub.microservices.ch06.attachment.data;

import com.packtpub.microservices.ch06.attachment.models.Attachment;
import org.springframework.data.repository.CrudRepository;

import java.util.List;

public interface AttachmentRepository extends
CrudRepository<Attachment, String> {
    public List<Attachment> findByMessageId(String messageId);
}
```

（5）我们还需要一种对传入请求建模的方式。我们的服务将接收请求正文中以 JSON 格式发送的请求。其中，JSON 对象将包含文件名和文件数据（文件数据是 Base64 编码的字符串）。对此使用以下定义创建一个名为 AttachmentRequest 的新类：

```
package com.packtpub.microservices.ch06.attachment.models;

import com.fasterxml.jackson.annotation.JsonProperty;
```

```java
import java.util.Map;

public class AttachmentRequest {
    private String fileName;

    private String data;

    public AttachmentRequest() {}

    public AttachmentRequest(String fileName, String data) {
        this.fileName = fileName;
        this.data = data;
    }

    public String getFileName() {
        return fileName;
    }

    public void setFileName(String fileName) {
        this.fileName = fileName;
    }

    public String getData() {
        return data;
    }

    public void setData(String data) {
        this.data = data;
    }

    @JsonProperty("file")
    private void unpackFileName(Map<String, String> file) {
        this.fileName = file.get("name");
        this.data = file.get("data");
    }
}
```

（6）在接下来将定义的控制器中，如果找不到特定消息的附件，则需要向调用方返回 HTTP 404 响应。为此，可创建一个名为 com.packtpub.microservices.ch06.attachment.exceptions 的新程序包和一个名为 AttachmentNotFoundException 的新类，具体如下：

```
package com.packtpub.microservices.ch06.attachment.exceptions;

import org.springframework.http.HttpStatus;
import org.springframework.web.bind.annotation.ResponseStatus;

@ResponseStatus(code = HttpStatus.NOT_FOUND, reason = "No
attachment(s) found")
public class AttachmentNotFoundException extends RuntimeException{}
```

（7）我们将所有内容放到控制器中。在此基本示例中，定义了两个方法：一个方法将列出特定消息的附件；另一个方法将创建新的附件。附件将被上传到 Amazon S3 存储桶中，其名称在配置值中被指定。对此，可创建一个名为 com.packtpub.microservices.ch06.attachment.controllers 的新程序包和一个名为 AttachmentController 的新类，具体如下：

```
package com.packtpub.microservices.ch06.attachment.controllers;

import com.amazonaws.services.s3.AmazonS3;
import com.amazonaws.services.s3.model.CannedAccessControlList;
import com.amazonaws.services.s3.model.ObjectMetadata;
import com.amazonaws.services.s3.model.PutObjectRequest;
import com.packtpub.microservices.ch06.attachment.data.
AttachmentRepository;
import com.packtpub.microservices.ch06.attachment.exceptions.
AttachmentNotFoundException;
import com.packtpub.microservices.ch06.attachment.models.Attachment;
import com.packtpub.microservices.ch06.attachment.models.
AttachmentRequest;
import org.apache.commons.codec.binary.Base64;
import org.springframework.beans.factory.annotation.Autowired;
import org.springframework.beans.factory.annotation.Value;
import org.springframework.web.bind.annotation.*;

import java.io.ByteArrayInputStream;
import java.io.InputStream;
import java.util.List;

@RestController
public class AttachmentController {

    @Autowired
    private AttachmentRepository attachmentRepository;

    @Autowired
```

```java
    private AmazonS3 s3Client;

    @Value("${s3.bucket-name}")
    private String bucketName;

    @RequestMapping(path = "/message/{message_id}/attachments",
method = RequestMethod.GET, produces = "application/json")
    public List<Attachment>
getAttachments(@PathVariable("message_id") String messageId) {
        List<Attachment> attachments =
attachmentRepository.findByMessageId(messageId);
        if (attachments.isEmpty()) {
            throw new AttachmentNotFoundException();
        }
        return attachments;
    }

    @RequestMapping(path = "/message/{message_id}/attachments",
method = RequestMethod.POST, produces = "application/json")
    public Attachment create(@PathVariable("message_id") String
messageId, @RequestBody AttachmentRequest request) {

        byte[] byteArray = Base64.decodeBase64(request.getData());

        ObjectMetadata metadata = new ObjectMetadata();
        metadata.setContentLength(byteArray.length);
        metadata.setContentType("image/jpeg");
        metadata.setCacheControl("public, max-age=31536000");
        InputStream stream = new ByteArrayInputStream(byteArray);

        String fullyResolved = String.format("%s/%s", messageId,
request.getFileName());

        s3Client.putObject(
            new PutObjectRequest(bucketName, fullyResolved, stream,
metadata)
.withCannedAcl(CannedAccessControlList.PublicRead));

        String url =
String.format("https://%s.s3.amazonaws.com/%s", bucketName,
fullyResolved);

        Attachment attachment = new Attachment(messageId, url,
```

```
request.getFileName(), 1);
        attachmentRepository.save(attachment);
        return attachment;
    }
}
```

（8）为了使其中任何一项都能起作用，我们必须创建一个属性文件。需要说明的是，Java 属性文件支持从环境变量中获取值的语法。据此，在 src/main/resources 目录中创建一个名为 application.yml 的新文件，具体如下：

```
spring:
    jpa.hibernate.ddl-auto: create
    datasource.url: ${DATABASE_URL}
    datasource.username: ${DATABASE_USERNAME}
    datasource.password: ${DATABASE_PASSWORD}

s3:
    bucket-name: ${BUCKET_NAME}
```

这个例子足够好用。AWS 开发工具包中的 EnvironmentVariableCredentialsProvider 期望设置 AWS_ACCESS_KEY_ID 和 AWS_SECRET_ACCESS_KEY，并且我们指定应该以类似方式从环境变量中读取多个非敏感配置值。这显然比对配置值进行硬编码更好，但是我们仍然向与服务在相同容器或进程空间中运行的任何进程公开了机密信息，还必须在某个地方设置环境变量（通过配置管理系统或在 Dockerfile 中指定），因此我们尚未解决存储敏感机密的问题。接下来，我们将修改新服务以从 Vault 中读取 S3 凭据。

提示：

在本秘笈中，我们将在开发模式下运行 Vault。安装 Vault 以用于生产环境是一个很宏大的主题，在单个秘笈中无法讨论完全。因此，有关 Vault 在生产环境中的使用信息，请查阅以下网址中提供的优秀文档：

https://www.vaultproject.io/intro/index.html

（9）在本地开发计算机上安装 Vault。有关各个平台上的安装指导，请访问以下网址：

http://www.vaultproject.io

如果你正在运行 macOS X 并使用 HomeBrew，则可以使用以下命令安装 Vault：

```
$ brew install vault
```

（10）在开发模式下运行 vault server，提供一个易于记忆的根令牌，命令如下：

```
$ vault server --dev --dev-root-token-
id="00000000-0000-0000-0000-000000000000"
```

（11）使用特定于此服务的路径启用 kv secrets 引擎的新实例，命令如下：

```
$ vault secrets enable -path=secret/attachment-service
```

（12）首先将 AWS 访问密钥和 secret 对作为机密信息写入 Vault 中，然后将占位符替换为实际的 AWS 访问密钥 ID 和 AWS 秘密访问密钥，命令如下：

```
$ vault write secret/attachment-service
attachment.awsAccessKeyId=<access-key>
attachment.awsSecretAccessKey=<access-secret>
```

（13）为了使我们的服务从 Vault 中读取这些值，可以使用一个库来简化 Spring Boot 应用程序的 Vault 集成。据此，修改项目的 build.gradle 文件并添加以下依赖项：

```
group 'com.packtpub.microservices'
version '1.0-SNAPSHOT'

buildscript {
    repositories {
        mavenCentral()
    }
    dependencies {
        classpath group: 'org.springframework.boot', name: 'spring-
boot-gradle-plugin', version: '1.5.9.RELEASE'
    }
}

apply plugin: 'java'
apply plugin: 'org.springframework.boot'
apply plugin: 'io.spring.dependency-management'

sourceCompatibility = 1.8

repositories {
    mavenCentral()
}

dependencies {
    compile group: 'org.springframework.boot', name: 'spring-boot-
starter-web'
    compile group: 'org.springframework.boot', name: 'spring-boot-
```

```
starter-data-jpa', version: '1.5.9.RELEASE'
    compile group: 'org.springframework.cloud', name: 'spring-
cloud-starter-vault-config', version: '1.1.1.RELEASE'
    compile group: 'mysql', name: 'mysql-connector-java'
    compile group: 'com.amazonaws', name: 'aws-java-sdk-s3',
version: '1.11.375'
    testCompile group: 'junit', name: 'junit', version: '4.12'
}
```

（14）我们的应用程序需要一个配置类来存储从 Vault 中读取的值。据此，创建一个名为 com.packtpub.microservices.ch06.attachment.config 的新程序包和一个名为 Configuration 的新类，具体如下：

```
package com.packtpub.microservices.ch06.attachment.config;

import org.springframework.boot.context.properties.
ConfigurationProperties;

@ConfigurationProperties("attachment")
public class Configuration {

    private String awsAccessKeyId;

    private String awsSecretAccessKey;

    public String getAwsAccessKeyId() {
        return awsAccessKeyId;
    }

    public void setAwsAccessKeyId(String awsAccessKeyId) {
        this.awsAccessKeyId = awsAccessKeyId;
    }

    public String getAwsSecretAccessKey() {
        return awsSecretAccessKey;
    }

    public void setAwsSecretAccessKey(String awsSecretAccessKey) {
        this.awsSecretAccessKey = awsSecretAccessKey;
    }
}
```

（15）修改 Application 类以创建我们刚刚创建的类的实例。然后在创建 S3 客户端时使用该实例，以便可以使用从 Vault 中获取的凭据，而不是环境变量，具体如下：

```java
package com.packtpub.microservices.ch06.attachment;

import com.amazonaws.auth.AWSCredentials;
import com.amazonaws.auth.AWSStaticCredentialsProvider;
import com.amazonaws.auth.BasicAWSCredentials;
import com.amazonaws.regions.Regions;
import com.amazonaws.services.s3.AmazonS3;
import com.amazonaws.services.s3.AmazonS3ClientBuilder;

import com.packtpub.microservices.ch06.attachment.config.Configuration;
import org.springframework.boot.SpringApplication;
import org.springframework.boot.autoconfigure.SpringBootApplication;
import org.springframework.boot.context.properties.EnableConfigurationProperties;
import org.springframework.context.annotation.Bean;

@SpringBootApplication
@EnableConfigurationProperties(Configuration.class)
public class Application {

    private final Configuration config;

    public Application(Configuration config) {
        this.config = config;
    }

    @Bean
    public AmazonS3 getS3Client() {
        AmazonS3ClientBuilder client = AmazonS3ClientBuilder.standard();
        AWSCredentials credentials = new BasicAWSCredentials(config.getAwsAccessKeyId(), config.getAwsSecretAccessKey());
        return client.withCredentials(
                new AWSStaticCredentialsProvider(credentials)).withRegion(Regions.US_WEST_2).build();
    }

    public static void main(String[] args) {
```

```
        SpringApplication.run(Application.class, args);
    }
}
```

至此,我们已将 attachment-service 服务配置为从 Vault 中读取 AWS 凭据。

6.5 安全日志记录

日志与跟踪和度量一起,构成了可观察系统的重要组成部分(本书第 7 章"监控和可观察性"将更广泛地讨论"可观察性")。

日志是源自特定系统中的有序且包含时间戳的事件序列。

在微服务架构中,拥有多种服务将导致复杂性日益增加,因此拥有良好的日志至关重要。什么是良好的日志?这并没有一个确切的客观标准,多数的讨论都是主观的,但总的来说,良好的日志应帮助工程师将可能导致特定错误状态或问题的事件组合在一起,使工程师能快速找出问题所在的原因。日志通常按级别进行组织,这是一个可配置的切换开关,允许开发人员指示服务发送到日志中的信息的多少。

日志对于观察生产环境中的系统行为至关重要,但日志也可能带来隐私和安全风险。若从系统发送过多的信息到日志中,则可能会使潜在的攻击者获得有关系统用户的信息或者敏感信息(如令牌或密钥),这些都可用于攻击系统的其他部分。拥有微服务架构可以分散并扩大这种可能的攻击面,因此针对服务制定周密的日志策略变得尤为重要。

6.6 基础架构即代码

微服务架构通常需要更频繁地配置计算资源。在系统中拥有更多节点会增加攻击面,攻击者可以扫描攻击面以寻找可能的漏洞。使系统易受攻击的最简单方法之一是失去对 Inventory 主机配置的跟踪,并使多个异构配置处于活动状态。

6.6.1 理论阐释

在配置管理系统(如 Puppet 或 Ansible)流行之前,通常有一组自定义的 shell 脚本,这些脚本会引导(Bootstrap)系统中的新服务器。这足够好用,但是随着系统需求的增长和 shell 脚本的修改,要让系统的陈旧部分适应不断变化的最新标准变得越来越难以为继。这种类型的配置漂移通常会使系统的遗留部分很容易受到攻击。配置管理通过允许团队使用代码来描述应如何配置系统中的节点（这些代码通常具有声明性语法），从而解决了许多此类问题。一般来说，配置管理系统不会去处理如何配置实际的计算资源（如计算节点、数据存储或网络存储）。

基础架构即代码（Infrastructure as Code，IaC）是通过机器可读的代码文件（而不是手动）管理基础架构配置和维护的过程。使用代码来描述基础结构可以有效地对系统进行版本控制、检查和回滚。它能够自动启动数据库节点，或将计算节点添加到集群中，使得开发人员不必担心应用程序，从而相对确保陈旧配置不会处于放任状态。基础架构即代码（IaC）与不变的基础架构结合在一起，可提供额外的安全网，以防止系统受到易受攻击的、被遗忘的组件的危害。

在本秘笈中，我们将演示如何使用由 HashiCorp 创建的开源工具 Terraform 来提供 AWS 资源的集合，包括 EC2 实例和 Redis ElastiCache。我们将保证配置有 Terraform 的资源在网络访问、备份和其他安全性方面共享配置。

6.6.2 实战操作

本秘笈需要执行以下操作。

（1）在使用 terraform 之前，必须先安装它。有关其安装方法的详细指导可在项目站点上找到，但是如果你正在运行 macOS X 并使用的是 HomeBrew（https://brew.sh/），则可以通过以下命令安装：

```
$ brew install terraform
```

（2）创建一个名为 example.tf 的新文件。这将包含我们的 EC2 实例和 ElastiCache 实例的配置。我们将使用默认的亚马逊机器镜像（Amazon Machine Image，AMI）并启用将保留 5 天的每日快照，具体如下：

```
provider "aws" {
    access_key = "ACCESS_KEY"
    secret_key = "SECRET_KEY"
    region     = "us-east-1"
}

resource "aws_instance" "example" {
    ami           = "ami-b374d5a5"
    instance_type = "t2.micro"
}

resource "aws_elasticache_cluster" "example" {
```

```
    cluster_id               = "cluster-example"
    engine                   = "redis"
    node_type                = "cache.m3.medium"
    num_cache_nodes          = 1
    parameter_group_name     = "default.redis3.2"
    port                     = 6379
    snapshot_window          = "05:00-09:00"
    snapshot_retention_limit = 5
}
```

注意将 ACCESS_KEY 和 SECRET_KEY 替换为有效的 AWS 访问密钥对。

（3）初始化 terraform。这将安装先前文件中引用的 AWS 提供程序，命令如下：

```
$ terraform init
```

（4）Terraform 的工作方式是，先提出一个执行计划，然后询问你是否要应用该计划。运行以下命令，并在出现提示时输入 yes：

```
$ terraform apply

aws_instance.example: Refreshing state... (ID: i-09b5cf5ed923d60f4)
```

执行计划已生成，并在以下代码中显示。资源操作将用以下符号表示：+create。Terraform 将执行以下操作：

```
+ aws_elasticache_cluster.example
      id: <computed>
      apply_immediately: <computed>
      availability_zone: <computed>
      az_mode: <computed>
      cache_nodes.#: <computed>
      cluster_address: <computed>
      cluster_id: "cluster-example"
      configuration_endpoint: <computed>
      engine: "redis"
      engine_version: <computed>
      maintenance_window: <computed>
      node_type: "cache.m3.medium"
      num_cache_nodes: "1"
      parameter_group_name: <computed>
      port: "6379"
      replication_group_id: <computed>
      security_group_ids.#: <computed>
      security_group_names.#: <computed>
```

```
            snapshot_retention_limit: "5"
            snapshot_window: "05:00-09:00"
            subnet_group_name: <computed>

Plan: 1 to add, 0 to change, 0 to destroy.
Do you want to perform these actions?
```

Terraform 将执行上述操作。你需要输入 yes 以批准操作，具体如下：

```
Enter a value: yes

...
```

（5）登录到 AWS 管理控制台，你将看到已创建一个新的 Redis 集群和 EC2 实例。Terraform 还可以帮助你清理。要销毁这两个资源，可以运行以下 destroy 命令并在出现提示时输入 yes：

```
$ terraform destroy
```

Terraform 是非常强大的工具。在此秘笈中，我们使用它创建了一个 EC2 实例和一个 ElastiCache 集群实例。开发人员可以使用 Terraform 执行更多的负载。

有关基础架构即代码（IaC）的主题内容非常丰富，它本身就足够被编写出一本厚书。我们建议你阅读由 HashiCorp 提供的说明文档，其网址如下：

https://www.terraform.io/docs/index.html

使用基础架构即代码（IaC）解决方案将使资源调配和管理更加安全，这样，即使你使用了过期配置，也不太可能失去对遗留基础架构的跟踪，从而强化了系统的安全性。

第 7 章 监控和可观察性

本章包含以下操作秘笈。
- 结构化 JSON 日志记录。
- 使用 StatsD 和 Graphite 收集度量值。
- 使用 Prometheus 收集度量值。
- 通过跟踪使调试更容易。
- 出现问题时发出警报。

7.1 导　语

微服务增加了架构的复杂性。随着系统中活动部件的增加，监控和观察系统行为变得越来越重要，也更具挑战性。在微服务架构中，影响一项服务的故障状况可能会以意想不到的方式级联传导，从而影响整个系统。数据中心某处的交换机故障可能会导致服务出现异常高的延迟，也可能导致源自 API 网关的请求间歇性超时，这些都可能会导致意外的用户影响，从而引发警报。诸如此类情况在微服务架构中并不罕见，所以需要预先考虑在内，以便工程师可以轻松确定影响客户端的事件的性质。如前文所述，分布式系统几乎必然会遇到某些故障，因此必须特别考虑将可观察性（Observability）构建到系统中。

微服务必须进行的另一项转变是转向开发运维一体化（DevOps）。许多传统的监控解决方案是在特殊的系统管理员或运营工程师组成的小组独自负责运营时开发的。系统管理员和运营工程师通常对系统级或主机级度量指标感兴趣，如 CPU、内存、磁盘和网络使用情况。这些指标很重要，但仅占可观察性的一小部分。编写微服务的工程师还必须考虑其他度量指标的可观察性，使用度量指标来观察系统特有的事件，例如抛出某些异常类型或发送到队列中的事件数量同样重要。

对可观察性进行规划还为我们提供了有效测试生产系统所需的信息。用于模拟和集成测试的临时环境可能会很有用，但是有些故障状态类别是无法测试的。如第 5 章"可靠性模式"所述，游戏日演习和其他形式的故障注入对于提高系统的弹性至关重要。可观察的系统适合进行此类测试，从而使工程师对理解系统充满信心。

本章将介绍若干个用于监控和可观察性的技术。我们将演示如何修改服务以发出结构化日志。我们还将研究度量指标，使用许多不同的系统来收集、聚合和可视化指标。最后，我们将讨论跟踪（Tracing），这是一种查看请求的方式，这些请求在系统的各个组件之间传播，并在检测到影响用户的错误情况时提醒我们。

7.2 结构化 JSON 日志记录

输出有用的日志是构建可观察服务的关键部分。什么是有用的日志？对于其构成内容的评判虽然是主观的，但是有一套很好的准则，即日志应包含有关系统中关键事件的时间戳信息。一个好的日志系统支持可配置日志级别（Configurable Log Level）的概念，因此可以根据使用该系统的工程师的需要，在特定的时间内拨入或拨出发送到日志中的信息量。例如，当针对生产环境中的失败场景测试服务时，可以提高日志级别并获取有关系统中事件的更多详细信息，这可能是非常有用的。

7.2.1 理论阐释

Java 应用程序中两个受欢迎的日志记录库是 Log4j 和 Logback，其网址分别如下：

https://logging.apache.org/log4j/2.x/

https://logback.qos.ch/

默认情况下，这两个库都将以非结构化格式发出日志条目，通常是用空格分隔的字段，其中包括时间戳、日志级别和消息等信息。这很有用，而且在微服务架构中尤其如此，因为在微服务架构中，可能会有多个服务将事件日志发送到集中式日志存储中。此外，发出某些一致性的结构化日志也非常有用。

JSON 已成为在系统之间传递消息的通用标准。几乎每种流行语言都具有用于解析和生成 JSON 的库。它很轻巧，但却是结构化的，这使其成为数据（如事件日志）的理想选择。发出 JSON 格式的事件日志将使你更轻松地将服务的日志输入中心存储中，并对日志数据进行分析和查询。

在本秘笈中，我们将使用流行的 Java 应用程序的 logback 库修改 message-service 服务以发出日志。

7.2.2 实战操作

本秘笈需要执行以下操作。

（1）打开第 6 章"安全性"中的 message-service 服务项目。我们要做的第一个更改是将 logback 库添加到 build.gradle 文件中，具体如下：

```
group 'com.packtpub.microservices'
version '1.0-SNAPSHOT'

buildscript {
    repositories {
        mavenCentral()
    }
    dependencies {
        classpath group: 'org.springframework.boot', name: 'spring-boot-gradle-plugin', version: '1.5.9.RELEASE'
    }
}

apply plugin: 'java'
apply plugin: 'org.springframework.boot'

sourceCompatibility = 1.8

repositories {
    mavenCentral()
}

dependencies {
    compile group: 'org.springframework.boot', name: 'spring-boot-starter-web'
    compile group: 'io.github.resilience4j', name: 'resilience4j-circuitbreaker', version: '0.11.0'
    compile group: 'net.logstash.logback', name: 'logstash-logback-encoder', version: '4.7'
    testCompile group: 'junit', name: 'junit', version: '4.12'
}
```

（2）创建一个 logback.xml 配置文件。在配置文件中，我们将创建一个名为 jsonLogger 的日志记录器，该日志记录器将引用一个名为 consoleAppender 的追加程序（Appender），具体如下：

```
<?xml version="1.0" encoding="utf-8"?>
<configuration>
    <appender name="consoleAppender"
```

```xml
        class="ch.qos.logback.core.ConsoleAppender">
        <encoder
class="net.logstash.logback.encoder.LogstashEncoder"/>
    </appender>
    <logger name="jsonLogger" additivity="false" level="DEBUG">
        <appender-ref ref="consoleAppender"/>
    </logger>
    <root level="INFO">
        <appender-ref ref="consoleAppender"/>
    </root>
</configuration>
```

（3）将单个示例日志消息添加到 Application.java 中，以测试我们的新日志记录配置，具体如下：

```java
package com.packtpub.microservices.ch07.message;

import com.packtpub.microservices.ch07.message.clients.SocialGraphClient;
import org.apache.log4j.LogManager;
import org.apache.log4j.Logger;
import org.springframework.boot.SpringApplication;
import org.springframework.boot.autoconfigure.SpringBootApplication;
import org.springframework.context.annotation.Bean;
import org.springframework.scheduling.annotation.EnableAsync;
import org.springframework.scheduling.concurrent.ThreadPoolTaskExecutor;

import java.util.concurrent.Executor;

@SpringBootApplication
@EnableAsync
public class Application {

    private Logger logger =
LogManager.getLogger(Application.class);

    @Bean
    public MessageRepository messageRepository() {
        return new MessageRepository();
    }

    @Bean
    public SocialGraphClient socialGraphClient() {
        return new SocialGraphClient("http://localhost:4567");
```

```
    }

    public static void main(String[] args) {
        logger.info("Starting application");
        SpringApplication.run(Application.class, args);
    }

    @Bean
    public Executor asyncExecutor() {
        ThreadPoolTaskExecutor executor = new ThreadPoolTaskExecutor();
        executor.setCorePoolSize(2);
        executor.setMaxPoolSize(2);
        executor.setQueueCapacity(500);
        executor.setThreadNamePrefix("SocialServiceCall-");
        executor.initialize();
        return executor;
    }
}
```

（4）运行应用程序，然后查看以 JSON 发出的日志消息，具体如下：

```
$ ./gradlew bootRun

> Task :bootRun
{"@timestamp":"2018-08-09T22:08:22.959-05:00","@version":1,"message":
"Starting
application","logger_name":"com.packtpub.microservices.ch07.message
.Application","thread_name":"main","level":"INFO","level_value":20000}

  .   ____          _            __ _ _
 /\\ / ___'_ __ _ _(_)_ __  __ _ \ \ \ \
( ( )\___ | '_ | '_| | '_ \/ _` | \ \ \ \
 \\/  ___)| |_)| | | | | || (_| |  ) ) ) )
  '  |____| .__|_| |_|_| |_\__, | / / / /
 =========|_|==============|___/=/_/_/_/
 :: Spring Boot ::        (v1.5.9.RELEASE)

{"@timestamp":"2018-08-09T22:08:23.786-05:00","@version":1,"message
":"Starting Application on fartlek.local with PID 82453
(/Users/posman/projects/microservices-cookbook/chapter07/message-
service/build/classes/java/main started by posman in
/Users/posman/projects/microservices-cookbook/chapter07/message-
```

service)","logger_name":"com.packtpub.microservices.ch07.message.Application","thread_name":"main","level":"INFO","level_value":20000}

7.3 使用 StatsD 和 Graphite 收集度量值

度量值（Metric）是一段时间内的数字度量。我们系统中收集的最常见的度量标准类型是计数器、计时器和仪表。

- 计数器（Counter）类型的确切含义是，值在某个时间段内递增的次数。
- 计时器（Timer）类型可用于测量系统中反复出现的事件，例如服务请求或执行数据库查询所花费的时间。
- 仪表（Gauge）类型是可以记录的任意数值。

7.3.1 理论阐释

StatsD 是 2011 年在 Etsy 发明的开源网络守护程序。度量数据通常被推到 statsd 服务器（该服务器通常在相同的服务器上）上，它将先聚合数据，然后将其发送到持久性后端。与 statsd 一起使用的最常见的后端之一是 Graphite，这是一种开源的时间序列存储引擎和绘图工具。Graphite 和 StatsD 共同构成了非常流行的度量堆栈。它们很容易上手，并拥有庞大的用户社区以及各种工具和库。

Spring Boot 有一个名为 Actuator 的子项目，该子项目为服务添加了许多可用于生产环境的功能。Actuator 免费为我们的服务提供了一些指标，并结合了一个名为 Micrometer 的项目。Actuator 支持与供应商无关的 API，可连接到各种度量指标的后端。在本秘笈和下一个秘笈中，我们将使用 Actuator 和 Micrometer。

在此秘笈中，我们将把 Actuator 添加到前面秘笈中修改过的 message-service 服务中。我们还将创建一些自定义的度量指标，并演示如何使用 statsd 和 graphite 绘制应用程序度量值的图形。我们将在 Docker 容器中以本地方式运行 statsd 和 graphite。

7.3.2 实战操作

本秘笈需要执行以下操作。

（1）打开前面秘笈中的 message-service 服务项目。我们将升级 Spring Boot 的版本，并将 Actuator 和 Micrometer 添加到依赖项列表中。据此，修改 build.gradle 文件，使其内容如下：

```
group 'com.packtpub.microservices'
version '1.0-SNAPSHOT'

buildscript {
    repositories {
        mavenCentral()
    }
    dependencies {
        classpath group: 'org.springframework.boot', name: 'spring-boot-gradle-plugin', version: '2.0.4.RELEASE'
    }
}

apply plugin: 'java'
apply plugin: 'org.springframework.boot'

sourceCompatibility = 1.8

repositories {
    mavenCentral()
}

dependencies {
    compile group: 'org.springframework.boot', name: 'spring-boot-starter-web', version: '2.0.4.RELEASE'
    compile group: 'org.springframework.boot', name: 'spring-boot-starter-actuator', version: '2.0.4.RELEASE'
    compile group: 'io.micrometer', name: 'micrometer-core', version: '1.0.6'
    compile group: 'io.micrometer', name: 'micrometer-registry-statsd', version: '1.0.6'
    compile group: 'io.github.resilience4j', name: 'resilience4j-circuitbreaker', version: '0.11.0'
    compile group: 'log4j', name: 'log4j', version: '1.2.17'
    compile group: 'net.logstash.logback', name: 'logstash-logback-encoder', version: '5.2'
    testCompile group: 'junit', name: 'junit', version: '4.12'
}
```

（2）在 src/main/resources 目录中打开 application.yml，并添加以下内容：

```
server:
    port:
```

```yaml
        8082
management:
  metrics:
    export:
      statsd:
        enabled: true
        flavor: "etsy"
        host:
          0.0.0.0
        port:
          8125
```

（3）我们的应用程序现在支持向本地运行的 statsd 实例发出指标。据此，打开 MessageController.java 并将 Timed 注解以及 get 方法添加到类中：

```java
package com.packtpub.microservices.ch07.message.controllers;

import com.packtpub.microservices.ch07.message.MessageRepository;
import com.packtpub.microservices.ch07.message.clients.SocialGraphClient;
import com.packtpub.microservices.ch07.message.exceptions.MessageNotFoundException;
import com.packtpub.microservices.ch07.message.exceptions.MessageSendForbiddenException;
import com.packtpub.microservices.ch07.message.models.Message;
import com.packtpub.microservices.ch07.message.models.UserFriendships;
import io.micrometer.core.annotation.Timed;
import io.micrometer.statsd.StatsdMeterRegistry;
import org.springframework.beans.factory.annotation.Autowired;
import org.springframework.http.ResponseEntity;
import org.springframework.scheduling.annotation.Async;
import org.springframework.web.bind.annotation.*;
import org.springframework.web.client.RestTemplate;
import org.springframework.web.servlet.support.ServletUriComponentsBuilder;

import java.net.URI;
import java.util.List;
import java.util.concurrent.CompletableFuture;

@RestController
@Timed
public class MessageController {
```

```java
    @Autowired
    private MessageRepository messagesStore;

    @Autowired
    private SocialGraphClient socialGraphClient;

    @Autowired
    private StatsdMeterRegistry registry;

    @Timed(value="get.messages")
    @RequestMapping(path = "/{id}", method = RequestMethod.GET,
produces = "application/json")
    public Message get(@PathVariable("id") String id) throws
MessageNotFoundException {
        registry.counter("get_messages").increment();
        return messagesStore.get(id);
    }

    @RequestMapping(path = "/", method = RequestMethod.POST,
produces = "application/json")
    public ResponseEntity<Message> send(@RequestBody Message
message) throws MessageSendForbiddenException {

        List<String> friendships =
socialGraphClient.getFriendships(message.getSender());
        if (!friendships.contains(message.getRecipient())) {
            throw new MessageSendForbiddenException("Must be
friends to send message");
        }

        Message saved = messagesStore.save(message);
        URI location = ServletUriComponentsBuilder
                .fromCurrentRequest().path("/{id}")
                .buildAndExpand(saved.getId()).toUri();
        return ResponseEntity.created(location).build();
    }

    @Async
    public CompletableFuture<Boolean> isFollowing(String fromUser,
String toUser) {
        String url = String.format(
```

```
"http://localhost:4567/followings?user=%s&filter=%s",
            fromUser, toUser);

    RestTemplate template = new RestTemplate();
    UserFriendships followings = template.getForObject(url,
UserFriendships.class);

    return CompletableFuture.completedFuture(
            followings.getFriendships().isEmpty()
    );
}
}
```

（4）为了证明度量值已经真正发出，我们将在 Docker 容器中以本地方式运行 statsd 和 graphite。在安装了 Docker 之后，运行以下命令，该命令将从 Docker Hub 取得镜像并以本地方式运行容器：

```
docker run -d --name graphite --restart=always \
  -p 80:80 -p 2003-2004:2003-2004 -p 2023-2024:2023-2024 \
  -p 8125:8125/udp -p 8126:8126 \
  hopsoft/graphite-statsd
```

（5）现在，可访问 http://localhost 以查看度量值。

7.4　使用 Prometheus 收集度量值

Prometheus 是一个开源监控和警报工具包，最初由 SoundCloud 于 2012 年开发。Prometheus 的灵感来自 Google 的 Borgmon。与诸如 statsd 之类的系统采用的推送（Push）模型相比，Prometheus 使用了拉取（Pull）模型来收集度量值。Prometheus 不是负责将度量值推送到 statsd 服务器，而是负责抓取具有度量值的服务所公开的端点。当大规模操作度量值时，这种责任倒置可以提供一些好处。

7.4.1　理论阐释

Prometheus 中的目标可以手动配置，也可以通过服务发现配置。

与诸如 Graphite 之类的系统用于存储度量数据的分层格式相反，Prometheus 采用了多维数据模型。Prometheus 中的时间序列数据由度量值名称（如 http_request_duration_seconds）和一个或多个标签（如 service = message-service 和 method = POST）标识。这种格式可以使跨多个不同应用程序的度量值标准化变得更加容易,这在微服务架构中特别有价值。

在本秘笈中,我们将继续使用 message-service 服务以及 Actuator 和 Micrometer 库。我们将配置 Micrometer 以使用 Prometheus 度量注册表,并公开 Prometheus 可以抓取的端点以便从我们的服务中收集指标。然后,我们将 Prometheus 配置为抓取 message-service 服务(在本地运行),并在本地运行 Prometheus,以验证我们可以查询度量值。

7.4.2 实战操作

本秘笈需要执行以下操作。

(1)打开 message-service 服务并编辑 build.gradle 文件,以使它包含 Actuator 和 Micrometer 的 Prometheus 依赖项,具体如下:

```
group 'com.packtpub.microservices'
version '1.0-SNAPSHOT'

buildscript {
    repositories {
        mavenCentral()
    }
    dependencies {
        classpath group: 'org.springframework.boot', name: 'spring-boot-gradle-plugin', version: '2.0.4.RELEASE'
    }
}

apply plugin: 'java'
apply plugin: 'org.springframework.boot'

sourceCompatibility = 1.8

repositories {
    mavenCentral()
}

dependencies {
    compile group: 'org.springframework.boot', name: 'spring-boot-starter-web', version: '2.0.4.RELEASE'
    compile group: 'org.springframework.boot', name: 'spring-boot-starter-actuator', version: '2.0.4.RELEASE'
```

```
    compile group: 'io.micrometer', name: 'micrometer-core',
version: '1.0.6'
    compile group: 'io.micrometer', name: 'micrometer-registry-
prometheus', version: '1.0.6'
    compile group: 'io.github.resilience4j', name: 'resilience4j-
circuitbreaker', version: '0.11.0'
    compile group: 'log4j', name: 'log4j', version: '1.2.17'
    compile group: 'net.logstash.logback', name: 'logstash-logback-
encoder', version: '5.2'
    testCompile group: 'junit', name: 'junit', version: '4.12'
}
```

（2）将以下内容添加到 application.yml 中，这将使公开度量值的端点能够收集到 Prometheus 度量指标注册表中（请注意，我们将为 actuator 添加的管理端点打开另一个端口）：

```
server:
    port:
        8082

management:
    server:
        port:
            8081
    endpoint:
        metrics:
            enabled: true
        prometheus:
            enabled: true
    endpoints:
        web:
            base-path: "/manage"
            exposure:
                include: "*"
    metrics:
        export:
            prometheus:
                enabled: true
```

（3）现在，我们可以测试服务是否在/manage/prometheus 端点上。对此，可运行服务并发出以下 curl 请求：

```
$ curl http://localhost:8081/manage/prometheus

# HELP tomcat_global_request_seconds
# TYPE tomcat_global_request_seconds summary
tomcat_global_request_seconds_count{name="http-nio-8082",} 0.0
tomcat_global_request_seconds_sum{name="http-nio-8082",} 0.0
# HELP tomcat_sessions_active_max
# TYPE tomcat_sessions_active_max gauge
tomcat_sessions_active_max 0.0
# HELP process_uptime_seconds The uptime of the Java virtual
machine
# TYPE process_uptime_seconds gauge
process_uptime_seconds 957.132
# HELP jvm_gc_live_data_size_bytes Size of old generation memory
pool after a full GC
# TYPE jvm_gc_live_data_size_bytes gauge
jvm_gc_live_data_size_bytes 1.9244032E7
```

（4）在 Docker 容器中配置并运行 Prometheus。在 /tmp 目录中创建一个名为 prometheus.yml 的新文件，其中包含有关目标的信息，具体如下：

```
# my global config
global:
    scrape_interval: 15s # Set the scrape interval to every 15
seconds. Default is every 1 minute.
    evaluation_interval: 15s # Evaluate rules every 15 seconds. The
default is every 1 minute.
    # scrape_timeout is set to the global default (10s).

# Alertmanager configuration
alerting:
    alertmanagers:
    - static_configs:
        - targets:
          # - alertmanager:9093

# Load rules once and periodically evaluate them according to the
global 'evaluation_interval'.
rule_files:
    # - "first_rules.yml"
    # - "second_rules.yml"
```

```yaml
# A scrape configuration containing exactly one endpoint to scrape:
# Here it's Prometheus itself.
scrape_configs:
    # The job name is added as a label `job=<job_name>` to any
timeseries scraped from this config.
    - job_name: 'prometheus'

      # metrics_path defaults to '/metrics'
      # scheme defaults to 'http'.

      static_configs:
      - targets: ['localhost:9090']

    - job_name: 'message-service'
      metrics_path: '/manage/prometheus'
      static_configs:
      - targets: ['localhost:8081']
```

（5）下载并提取适用于你平台的 Prometheus 版本。有关安装说明请访问 Prometheus 网站，其网址如下：

https://prometheus.io/docs/introduction/first_steps/

使用我们在步骤（4）中创建的配置文件运行 Prometheus，命令如下：

```
$ ./prometheus --config.file=/tmp/prometheus.yml
```

（6）在浏览器中打开 http:// localhost:9090，发出 Prometheus 查询并查看度量值。在开始向服务发出请求之前，你看到的度量值将只有 Java 虚拟机（JVM）和系统度量值，但这应该使你对可以使用 Prometheus 进行的查询类型有所了解，并演示了抓取程序的工作方式。

7.5 通过跟踪使调试更容易

在微服务架构中，单个请求可能会通过多个不同的服务，并导致写入多个不同的数据存储和事件队列中。在调试生产环境中的事件时，并不总是清楚一个系统或另一个系统中是否存在问题。缺乏特异性意味着，度量指标和日志仅构成生产图景的一小部分。

7.5.1 理论阐释

有时我们需要缩小并查看从用户代理到终端服务的请求的整个生命周期图景，然后返回并查看细节。

2010 年，Google 的工程师发表了一篇描述 Dapper 的论文，其网址如下：

https://research.google.com/archive/papers/dapper-2010-1.pdf

Dapper 是一种大规模的跟踪基础结构的分布式系统。上述论文描述了 Google 如何使用内部开发的跟踪系统来帮助观察系统行为和调试性能问题。这项工作启发了其他人，包括 Twitter 的工程师，他们在 2012 年推出了一种名为 Zipkin 的开源分布式跟踪系统。有关 Zipkin 的详细介绍可访问以下网址：

https://blog.twitter.com/engineering/en_us/a/2012/distributed-systems-tracing-with-zipkin.html

Zipkin 最初是 Dapper 论文的一种实现，但后来演变为一整套用于分析性能和检查对 Twitter 基础结构的请求的工具。

跟踪空间中正在进行的所有工作显然都需要某种标准化的 API。OpenTracing 框架就试图做到这一点。有关 OpenTracing 框架的详细介绍可访问以下网址：

http://opentracing.io/

OpenTracing 定义了一个规范，详细说明了跟踪的泛语言标准。来自不同公司的许多工程师为这项工作做出了贡献，其中包括最初创建了 Jaeger 的 Uber 工程师，相关说明可访问以下网址：

https://eng.uber.com/distributed-tracing/

Jaeger 是一种开源的、端到端的分布式跟踪系统，它符合 OpenTracing 规范。

在本秘笈中，我们将修改 message-service 服务以添加对跟踪的支持。然后，我们将在 Docker 容器中运行 Jaeger，以便在实践中看到一些跟踪结果。

7.5.2 实战操作

本秘笈需要执行以下操作。

（1）打开 message-service 项目，并用以下内容替换 build.gradle 的内容：

```
group 'com.packtpub.microservices'
version '1.0-SNAPSHOT'
```

```
buildscript {
    repositories {
        mavenCentral()
    }
    dependencies {
        classpath group: 'org.springframework.boot', name: 'spring-boot-gradle-plugin', version: '2.0.4.RELEASE'
    }
}

apply plugin: 'java'
apply plugin: 'org.springframework.boot'

sourceCompatibility = 1.8

repositories {
    mavenCentral()
}

dependencies {
    compile group: 'org.springframework.boot', name: 'spring-boot-starter-web', version: '2.0.4.RELEASE'
    compile group: 'org.springframework.boot', name: 'spring-boot-starter-actuator', version: '2.0.4.RELEASE'
    compile group: 'io.micrometer', name: 'micrometer-core', version: '1.0.6'
    compile group: 'io.micrometer', name: 'micrometer-registry-statsd', version: '1.0.6'
    compile group: 'io.opentracing.contrib', name: 'opentracing-spring-cloud-starter-jaeger', version: '0.1.13'
    compile group: 'io.github.resilience4j', name: 'resilience4j-circuitbreaker', version: '0.11.0'
    compile group: 'log4j', name: 'log4j', version: '1.2.17'
    compile group: 'net.logstash.logback', name: 'logstash-logback-encoder', version: '5.2'
    testCompile group: 'junit', name: 'junit', version: '4.12'
}
```

（2）在 src/main/resources 目录中打开 application.yml，并添加一个用于 opentracing 配置的部分。在本示例中，我们将配置 opentracing 实现，以连接到在端口 6831 上以本地方式运行的 Jaeger 实例，具体如下：

```
opentracing:
    jaeger:
        udp-sender:
            host: "localhost"
            port:
                6831
spring:
    application:
        name: "message-service"
```

（3）为了收集跟踪数据，我们将在本地运行 Jaeger 的实例。对此，可运行以下 Docker 命令：

```
docker run -d --name jaeger \
    -e COLLECTOR_ZIPKIN_HTTP_PORT=9411 \
    -p 5775:5775/udp \
    -p 6831:6831/udp \
    -p 6832:6832/udp \
    -p 5778:5778 \
    -p 16686:16686 \
    -p 14268:14268 \
    -p 9411:9411 \
    jaegertracing/all-in-one:latest
```

（4）运行 message-service 服务并提出一些示例请求（即使它们会导致 404）。在浏览器中打开 http://localhost:16686，你将看到 Jaeger 的 Web 用户界面。单击 Search（搜索），即可浏览到目前为止所收集到的跟踪数据。

7.6 出现问题时发出警报

当你深入研究微服务时会发现，很多微服务都可能被要求每天 24 小时乘以每周 7 天的全天候服务。客户将要求你的服务随时可用。一方面是，这一现实的可用性需求的增长；另一方面是，分布式系统不断出现的某种故障。因此，可以说没有任何系统是完全健康的。

无论你是采用单体架构还是微服务架构，尝试完全避免生产事故都是毫无意义的。相反，你能够做的应该是尝试优化故障响应能力，通过减少解决故障的时间来限缩故障对客户的不良影响。

7.6.1 理论阐释

减少解决事件所花费的时间通常以平均解决时间（Mean Time To Resolve，MTTR）来衡量，这首先就涉及减少平均检测时间（Mean Time To Detect，MTTD）。当服务处于影响客户的故障状态时，能够准确地给值班工程师发出警报对于维持正常运行时间至关重要。良好的警报应该是需要采取行动和紧急的；如果你的系统在故障不需要采取行动或不紧急（不影响客户）时通知值班工程师，则可能会让值班工程师不胜其烦并造成通常所说的警报疲劳。警报疲劳是非常真实的，比任何数量的软件错误或硬件故障对正常运行时间的灾难性影响更大。因此，必须不断改进系统的警报，以使阈值和其他因素恰到好处，以防止误报，同时针对真正影响客户的事件保持警报。

你不需要构建自己的警报基础结构，在这方面也有工具可以选择。例如，PagerDuty 就是一种软件即服务（Software as a Service，SaaS）工具，可让你为值守特定服务的工程师团队创建上报策略和时间表。使用 PagerDuty，你可以设置轮换时间表，例如，一个由 5 人组成的团队，其工程师预计每 5 个星期值守一个星期。上报策略允许你配置一组步骤，以防出现值班工程师不可用的状态（例如，他们也许正在高速公路上驾驶汽车）。上报策略通常配置为如果事件在一定时间内未被确认，则呼叫第二个值班工程师、经理，甚至整个团队。使用诸如 PagerDuty 之类的系统，可使团队中的工程师享受到更多不必值守的自由时光，同时又能迅速响应影响客户的事件。

可以使用任何数量的支持集成来手动配置警报，但这既耗时又容易出错。相反，最好有一个系统可以使你自动创建和维护服务警报。本章介绍的 Prometheus 监视和警报工具包即包括了一个名为 Alertmanager 的工具，使你可以做到这一点。在本秘笈中，我们将修改 message-service 服务以使用 Alertmanager 添加警报。具体来说，我们将配置一个警报，该警报将在平均响应时间超过 500ms 并且这种状况至少持续 5min 时触发。我们将从已经包含 Prometheus 度量指标的 message-service 服务版本开始工作。我们不会在此秘笈中添加任何 PagerDuty 集成，因为这将需要一个 PagerDuty 账户才能继续。PagerDuty 在其网站上有很详细的集成指南。我们将配置 Alertmanager 发送一个基于 WebHook 的简单警报。

7.6.2 实战操作

本秘笈需要执行以下操作。

（1）在先前的秘笈中使用名为 prometheus.yml 的文件配置了 Prometheus。现在我们需要将 alertmanagers 配置添加到此文件中，因此再次打开它并添加以下内容：

```yaml
# my global config
global:
    scrape_interval:      15s # Set the scrape interval to every 15 seconds. Default is every 1 minute.
    evaluation_interval:  15s # Evaluate rules every 15 seconds. The default is every 1 minute.
    # scrape_timeout is set to the global default (10s).

# alertmanagers configuration
alerting:
    alertmanagers:
    - static_configs:
        - targets:
            - localhost:9093

# Load rules once and periodically evaluate them according to the global 'evaluation_interval'.
rule_files:
    - "rules.yml"
  # - "first_rules.yml"
  # - "second_rules.yml"

# A scrape configuration containing exactly one endpoint to scrape:
# Here it's Prometheus itself.
scrape_configs:
  # The job name is added as a label `job=<job_name>` to any timeseries scraped from this config.
  - job_name: 'prometheus'

    # metrics_path defaults to '/metrics'
    # scheme defaults to 'http'.

    static_configs:
    - targets: ['localhost:9090']

  - job_name: 'message-service'
    metrics_path: '/manage/prometheus'
    static_configs:
    - targets: ['localhost:8081']
```

（2）创建一个名为/tmp/rules.yml 的新文件。该文件定义了我们希望 Prometheus 能够为以下情形创建警报的规则：

```yaml
groups:
 - name: message-service-latency
   rules:
    - alert: HighLatency
      expr: rate(http_server_requests_seconds_sum{job="message-service", instance="localhost:8081"}[1m]) / rate(http_server_requests_seconds_count{job="message-service", instance="localhost:8081"}[1m]) > .5
      for: 1m
      labels:
          severity: 'critical'
      annotations:
          summary: High request latency
```

（3）创建一个名为/tmp/alertmanager.yml 的新文件。这是将描述我们的警报配置的文件。它分为几个不同的部分，某些配置选项的全局集合会影响 alertmanager 的工作方式。称为 receivers（接收器）的部分就是我们配置警报通知系统的地方。在本示例中，它是以本地方式运行的服务的 WebHook。出于演示目的，我们将仅编写一个很小的 Ruby 脚本，侦听 HTTP 请求并将有效负载打印到标准输出中，具体如下：

```yaml
global:
    resolve_timeout: 5m

route:
    group_by: ['alertname']
    group_wait: 10s
    group_interval: 10s
    repeat_interval: 1h
    receiver: 'web.hook'

receivers:
 - name: 'web.hook'
   webhook_configs:
    - url: 'http://127.0.0.1:4567/'
```

（4）以下是很小的 Ruby 服务的源代码，它将输出警报：

```ruby
require 'sinatra'

post '/' do
    body = request.body.read()
    puts body
```

```
    return body
end
```

（5）运行 Ruby 脚本，重新启动 prometheus，然后启动 alertmanager。随着这 3 个系统的运行，我们即做好了测试警报的准备。对应的命令如下：

```
$ ruby echo.rb
...
$ ./prometheus --config.file=/tmp/prometheus.yml

$ ./alertmanager --config.file=/tmp/alertmanager.yml
...
```

（6）为了使警报启动，可以打开 message-service 服务并将以下行添加到 MessageController.java 中。这是单行代码，它将强制控制器在返回响应之前休眠 600ms。请注意，这高于规则配置中描述的 500ms 的阈值，具体如下：

```
@RequestMapping(path = "/{id}", method = RequestMethod.GET,
produces = "application/json")
public Message get(@PathVariable("id") String id) throws
MessageNotFoundException {
    try { Thread.sleep(600); } catch (InterruptedException e) }
e.printStackTrace(); }
    return messagesStore.get(id);
}
```

（7）在完成修改之后，即可运行更新的 message-service 服务并向其发出许多请求。5min 后，Prometheus 应通知 Alertmanager，后者应随后通知你调试 Ruby 服务。如果这一切发生，则证明你的警报是正常有效的。

第 8 章 扩 展

本章包含以下操作秘笈。
- 使用 Vegeta 对微服务进行负载测试。
- 使用 Gatling 对微服务进行负载测试。
- 构建自动扩展集群。

8.1 导 语

使用微服务而不是单体架构的一个显著优势是,微服务可以单独扩展以满足它们所服务的独特流量需求。与仅需要为特定种类的请求执行工作的服务相比,微服务必须对每个请求工作的服务具有截然不同的扩展需求。

由于微服务封装了对单个域实体的所有权,因此可以对它们进行独立的负载测试。它们也可以配置为根据需求自动缩放。本章将讨论使用两种不同的负载测试工具进行负载测试,以及在 AWS 中设置可以按需扩展的自动扩展组。最后,我们还将讨论容量规划策略。

8.2 使用 Vegeta 对微服务进行负载测试

负载测试是预测服务随着时间的推移将如何运行的重要部分。在执行负载测试时,我们不应该仅仅问一些简单的问题,例如"我们的系统每秒能够处理多少个请求?"相反,我们应该尝试了解整个系统在各种负载条件下的性能。为了回答这个问题,我们需要了解构成系统的基础结构以及特定服务所具有的依赖关系。

例如,服务是在负载均衡器背后吗?使用了内容分发网络(Content Delivery Network,CDN)吗?使用了其他缓存机制吗?如果系统具有良好的可观察性,则可以回答所有这些问题以及更多问题。

8.2.1 理论阐释

Vegeta 是一个开源负载测试实用程序,旨在以恒定的请求速率测试 HTTP 服务。它是一种多功能工具，可以用作命令行实用程序或库。在本秘笈中，我们将重点介绍使用命令行实用程序。通过 Vegeta，你可以将目标指定为单独文件中的 URL（可以选择使用自定义标头和请求主体），这些文件可用作命令行工具的输入。然后，命令行工具可以使用各种选项来控制请求速率和持续时间以及其他变量，从而攻击文件中的目标。

本秘笈将使用 Vegeta 来测试我们在前几章中一直使用的 message-service 服务。我们将测试一个简单的请求路径，其中包括创建新消息和检索消息列表。

8.2.2 实战操作

本秘笈需要执行以下操作。

（1）我们将修改 message-service 服务，并添加一个新的端点，该端点使我们能够查询特定用户的所有消息。这引入了收件箱（Inbox）的概念，因此我们将修改 MessageRepository 类，以将用户名的新内存中映射（In-Memory Map）添加到消息列表中。请注意，在实际生产系统中，应该选择一个更持久和灵活的存储方式，本示例中的内存中映射仅用于演示目的。本步骤对应的代码如下：

```java
package com.packtpub.microservices.ch08.message;

import com.packtpub.microservices.ch08.message.exceptions.MessageNotFoundException;
import com.packtpub.microservices.ch08.message.models.Message;

import java.util.*;
import java.util.concurrent.ConcurrentHashMap;

public class MessageRepository {

    private ConcurrentHashMap<String, Message> messages;
    private ConcurrentHashMap<String, List<Message>> inbox;

    public MessageRepository() {
        messages = new ConcurrentHashMap<>();
        inbox = new ConcurrentHashMap<>();
    }

    public Message save(Message message) {
        UUID uuid = UUID.randomUUID();
        Message saved = new Message(uuid.toString(),
message.getSender(), message.getRecipient(),
```

```
            message.getBody(), message.getAttachmentUri());
        messages.put(uuid.toString(), saved);
        List<Message> userInbox =
inbox.getOrDefault(message.getRecipient(), new ArrayList<>());
        userInbox.add(saved);
        inbox.put(message.getRecipient(), userInbox);
        return saved;
    }

    public Message get(String id) throws MessageNotFoundException {
        if (messages.containsKey(id)) {
            return messages.get(id);
        } else {
            throw new MessageNotFoundException("Message " + id + " could not be found");
        }
    }

    public List<Message> getByUser(String userId) {
        return inbox.getOrDefault(userId, new ArrayList<>());
    }
}
```

（2）修改 MessageController 以添加端点本身，具体如下：

```
package com.packtpub.microservices.ch08.message.controllers;

import com.packtpub.microservices.ch08.message.MessageRepository;
import com.packtpub.microservices.ch08.message.clients.SocialGraphClient;
import com.packtpub.microservices.ch08.message.exceptions.MessageNotFoundException;
import com.packtpub.microservices.ch08.message.exceptions.MessageSendForbiddenException;
import com.packtpub.microservices.ch08.message.exceptions.MessagesNotFoundException;
import com.packtpub.microservices.ch08.message.models.Message;
import com.packtpub.microservices.ch08.message.models.UserFriendships;
import org.springframework.beans.factory.annotation.Autowired;
import org.springframework.http.ResponseEntity;
import org.springframework.scheduling.annotation.Async;
import org.springframework.web.bind.annotation.*;
import org.springframework.web.client.RestTemplate;
import org.springframework.web.servlet.support.
```

```java
ServletUriComponentsBuilder;

import java.net.URI;
import java.util.List;
import java.util.concurrent.CompletableFuture;

@RestController
public class MessageController {

    @Autowired
    private MessageRepository messagesStore;

    @Autowired
    private SocialGraphClient socialGraphClient;

    @RequestMapping(path = "/{id}", method = RequestMethod.GET,
produces = "application/json")
    public Message get(@PathVariable("id") String id) throws
MessageNotFoundException {
        return messagesStore.get(id);
    }

    @RequestMapping(path = "/", method = RequestMethod.POST,
produces = "application/json")
    public ResponseEntity<Message> send(@RequestBody Message
message) throws MessageSendForbiddenException {
        List<String> friendships =
socialGraphClient.getFriendships(message.getSender());

        if (!friendships.contains(message.getRecipient())) {
            throw new MessageSendForbiddenException("Must be
friends to send message");
        }

        Message saved = messagesStore.save(message);
        URI location = ServletUriComponentsBuilder
                .fromCurrentRequest().path("/{id}")
                .buildAndExpand(saved.getId()).toUri();
        return ResponseEntity.created(location).build();
    }

    @RequestMapping(path = "/user/{userId}", method =
```

```
RequestMethod.GET, produces = "application/json")
    public ResponseEntity<List<Message>>
getByUser(@PathVariable("userId") String userId) throws
MessageNotFoundException {
        List<Message> inbox = messagesStore.getByUser(userId);
        if (inbox.isEmpty()) {
            throw new MessageNotFoundException("No messages found
for user: " + userId);
        }
        return ResponseEntity.ok(inbox);
    }

    @Async
    public CompletableFuture<Boolean> isFollowing(String fromUser,
String toUser) {
        String url = String.format(
"http://localhost:4567/followings?user=%s&filter=%s",
                fromUser, toUser);

        RestTemplate template = new RestTemplate();
        UserFriendships followings = template.getForObject(url,
UserFriendships.class);

        return CompletableFuture.completedFuture(
                followings.getFriendships().isEmpty()
        );
    }
}
```

（3）我们还需要一个模拟的 social-graph-service 服务，因此可以在一个名为 socialgraph.rb 的文件中创建以下 Ruby 脚本并运行它：

```
require 'sinatra'

get '/friendships/:user' do
    content_type :json
    {
        username: "user:32134",
        friendships: [
            "user:12345"
        ]
    }.to_json
end
```

（4）安装 vegeta。如果你使用的是 macOS X，并且已安装 HomeBrew，则可以使用以下方法：

```
$ brew update && brew install vegeta
```

（5）在使用 vegeta 之前，我们还需要创建一个 targets 文件。发出的第一个请求将使用指定的请求正文创建一条消息；第二个请求将按用户 ID 获取消息列表。对此，创建一个名为 message-request-body.json 的文件，具体如下：

```
{
    "sender": "user:32134",
    "recipient": "user:12345",
    "body": "Hello there!",
    "attachment_uri": "http://foo.com/image.png"
}
```

（6）创建另一个名为 targets.txt 的文件，具体如下：

```
POST http://localhost:8082/
Content-Type: application/json
@message-request-body.json

GET http://localhost:8082/user:12345
```

（7）在运行 message-service 服务和模拟 social-graph-service 服务的同时，即可使用以下代码对这两个服务进行负载测试：

```
$ cat targets.txt| vegeta attack -duration=60s -rate=100 | vegeta
report -reporter=text

Requests        [total, rate]               6000, 100.01
Duration        [total, attack, wait]       1m0.004668981s,
59.99172349s,   12.945491ms
Latencies       [mean, 50, 95, 99, max]     10.683968ms, 5.598656ms,
35.108562ms, 98.290388ms, 425.186942ms
Bytes In        [total, mean]               667057195, 111176.20
Bytes Out       [total, mean]               420000, 70.00
Success         [ratio]                     99.80%
Status Codes    [code:count]                201:3000    500:12  200:2988
Error Set:
50
```

可以使用不同的持续时间值和请求速率进行实验，以查看系统行为的变化方式。例如，如果将速率提高到 1000，会发生什么情况？根据硬件和其他因素，单线程 Ruby 模拟服务可能会不堪重负，并使添加到 message-service 服务中的断路器跳闸。这应该会更改某些细节，例如成功率，因此这是一个重要的观察点。

如果单独对模拟 Ruby 服务进行负载测试，会发生什么情况？

在此秘笈中，我们对 message-service 服务进行了负载测试，该服务取决于 social-graph-service 服务。两种服务都在本地运行，这对于演示目的来说是必要的，它使我们对这两个系统的行为有一些了解。在生产系统中，至关重要的是对生产中的服务进行负载测试，以便将服务请求所涉及的所有基础结构（如负载均衡器、缓存等）都涵盖其中。在生产系统中，你还可以监控仪表板并查看系统在负载条件下行为的变化方式。

8.3 使用 Gatling 对微服务进行负载测试

Gatling 是一种开源负载测试工具，允许用户使用基于 Scala 的 DSL 编写自定义方案的脚本。其测试方案不仅可以是简单的直线路径测试，也可以涉及多个步骤，甚至还可以模拟用户行为，例如暂停并根据测试中的输出来决定如何进行操作。Gatling 可用于自动化微服务，甚至是基于浏览器的 Web 应用程序的负载测试。

8.3.1 理论阐释

在前面的秘笈中，我们使用了 Vegeta 以恒定速率向 message-service 服务发送请求。我们的请求路径创建了一条新消息，然后检索了某个用户的所有消息。该方法的优势在于，随着消息列表的增加，能够测试检索用户的所有消息的响应时间。Vegeta 在这种类型的测试上表现很出色，但由于它是通过静态文件馈送攻击目标的，因此不能使用 Vegeta 根据先前请求的响应来构建动态请求路径。

由于 Gatling 使用 DSL 编写负载测试方案的脚本，因此可以发出请求，捕获响应的某些元素，并使用该输出来做出有关未来请求的决策。

在本秘笈中，我们将使用 Gatling 编写负载测试方案的脚本，该方案包括创建一条消息，然后通过其 ID 检索该特定消息。与我们之前的秘笈相比，这是一种迥然不同的测试，因此这是一个很好的证明 Vegeta 和 Gatling 之间差异的机会。

8.3.2 实战操作

本秘笈需要执行以下操作。

（1）首先需要为你的操作系统平台下载 Gatling。Gatling 以 ZIP 捆绑包的形式分发，可从以下地址下载安装文件：

https://gatling.io/download/

在下载完成之后，解压缩到你选择的目录中：

```
$ unzip gatling-charts-highcharts-bundle-2.3.1-bundle.zip
...
$ cd gatling-charts-highcharts-bundle-2.3.1
```

（2）默认情况下，将用于 Gatling 的模拟放置在 user-files/simulations 目录中。创建一个名为 messageservice 的新子目录和一个名为 BasicSimulation.scala 的新文件。这是包含描述你的方案的代码的文件。在我们的方案中，将使用 Gatling DSL 脚本编写一个 POST 请求，发送到创建消息的端点，然后编写一个 GET 请求，同样发送到消息端点，具体代码如下：

```scala
package messageservice

import io.gatling.core.Predef._
import io.gatling.http.Predef._
import scala.concurrent.duration._

class BasicSimulation extends Simulation {

    val httpConf = http
        .baseURL("http://localhost:8082")
        .acceptHeader("application/json")

    val scn = scenario("Create a message")
        .exec(
            http("createMessage")
                .post("/")
                .header("Content-Type", "application/json")
                .body(StringBody("""{"sender": "user:32134", "recipient": "user:12345", "body": "Hello there!", "attachment_uri": "http://foo.com/image.png"}"""))
                .check(header(HttpHeaderNames.Location).saveAs("location"))
        )
        .pause(1)
        .exec(
            http("getMessage")
```

第 8 章 扩　　展

```
        .get("${location}")
    )

    setUp(scn.inject(atOnceUsers(50)).protocols(httpConf))
}
```

（3）创建与上一个秘笈相同的模拟 Ruby 服务并运行它，具体如下：

```
require 'sinatra'

get '/friendships/:user' do
    content_type :json
    {
        username: "user:32134",
        friendships: [
            "user:12345"
        ]
    }.to_json
end
```

（4）运行 Ruby 模拟服务以及 message-service 服务。在 Gatling 目录中，通过运行 bin/gatling.sh 启动 Gatling。系统将提示你选择要运行的模拟，此时可以选择 messageservice.BasicSimulation，具体如下：

```
$ bin/gatling.sh
GATLING_HOME is set to /Users/posman/projects/microservices-
cookbook/chapter08/gatling-charts-highcharts-bundle-2.3.1
Choose a simulation number:
        [0] computerdatabase.BasicSimulation
        [1] computerdatabase.advanced.AdvancedSimulationStep01
        [2] computerdatabase.advanced.AdvancedSimulationStep02
        [3] computerdatabase.advanced.AdvancedSimulationStep03
        [4] computerdatabase.advanced.AdvancedSimulationStep04
        [5] computerdatabase.advanced.AdvancedSimulationStep05
        [6] messageservice.BasicSimulation
6
Select simulation id (default is 'basicsimulation'). Accepted
characters are a-z, A-Z, 0-9, - and _

Select run description (optional)

Simulation messageservice.BasicSimulation started...
...
```

（5）此时，输出将显示有关负载测试结果的一些统计信息。请求的响应时间将被分类为 800ms 以下、800ms～1200ms 以及 1200ms 以上。输出结果中还将显示一个指向 HTML 文件的链接。单击该链接，即可在浏览器中打开它，进而查看有关负载测试的图表和其他有用的可视化结果。

正如我们在本秘笈中所看到的，Gatling 在运行负载测试方面提供了很多灵活性。通过使用 DSL 的一些巧妙脚本，可以更好地模拟生产环境中的流量，解析日志文件和生成请求，基于等待时间、响应或其他请求元素做出动态决策。Gatling 和 Vegeta 都是出色的负载测试工具，可用于探索系统在各种负载条件下的运行方式。

8.4 构建自动扩展集群

随着虚拟化（Virtualization）的出现以及向基于云的基础设施的转移，应用程序现在已经可以存在于弹性基础设施上，这些基础设施被设计为可以根据预期或测量的流量模式进行增长和收缩。如果你的应用程序已经经过了高峰期，那么就不必在非高峰期也调配全部容量，从而浪费了计算资源和金钱。从虚拟化到容器和容器调度程序，采用不断变化的动态基础设施以适应系统需求的情况现在越来越普遍。

8.4.1 理论阐释

微服务非常适合自动扩展。因为我们可以分别扩展系统的各个部分，所以更容易衡量特定服务及其依赖项的扩展需求。

有很多种方法都可以创建自动扩展集群。在第 9 章中，我们将讨论容器编排工具，但本节也无须跳过，因为我们可以在任何云提供商的程序中创建自动扩展集群。在本秘笈中，我们将介绍使用 Amazon Web Services（特别是 Amazon EC2 Auto Scaling）创建自动扩展计算集群。我们将创建一个集群，其中有多个 EC2 实例在应用程序负载均衡器（Application Load Balancer，ALB）后运行 message-service 服务。我们将配置集群以根据 CPU 利用率的情况自动添加实例。

8.4.2 实战操作

本秘笈需要执行以下操作。

（1）本操作秘笈需要一个 AWS 账户。如果你还没有 AWS 账户，则可以访问以下链接创建一个账户：

https://aws.amazon.com/premiumsupport/knowledge-center/create-and-activate-aws-account/

然后在以下位置创建一组访问密钥：

https://docs.aws.amazon.com/general/latest/gr/managing-aws-access-keys.html

现在安装 aws cli 实用程序。如果你使用的是 macOS X，并且已安装 HomeBrew，则可以执行以下操作：

```
$ brew install aws
```

（2）配置 aws 命令行实用程序，输入你创建的访问密钥：

```
$ aws configure
```

（3）创建启动配置。启动配置是创建新实例时自动伸缩组使用的模板。在本示例中，我们选择了 Amazon AMI 和 t2.nano 作为 EC2 实例类型。有关 EC2 实例类型的详细说明可访问以下网址：

https://aws.amazon.com/ec2/instance-types/

创建启动配置的代码如下：

```
$ aws autoscaling create-launch-configuration --launch-
configuration-name message-service-launch-configuration --image-id
ari-f606f39f --instance-type t2.nano
```

（4）创建实际的自动缩放组。自动缩放组具有可配置的最大和最小设置，这些大小指定自动缩放组可以根据需求缩小或增长多少。在本示例中，我们将创建一个最少包含 1 个实例，最多包含 5 个实例的自动缩放组，如以下代码所示：

```
$ aws autoscaling create-auto-scaling-group --auto-scaling-group-
name message-service-asg --launch-configuration-name message-
service-launch-configuration --max-size 5 --min-size 1 --
availability-zones "us-east-1a"
```

（5）我们希望自动伸缩组中的实例可以在负载均衡器后面访问，因此先要创建一个负载均衡器，代码如下：

```
$ aws elb create-load-balancer --load-balancer-name message-
service-lb --listeners
"Protocol=HTTP,LoadBalancerPort=80,InstanceProtocol=HTTP,
InstancePort=8082" --availability-zones us-east-1a
```

```
{
    "DNSName": "message-service-lb-1741394248.us-east-1.elb.amazonaws.com"
}
```

（6）为了自动缩放我们的自动缩放组，还需要定义一个指标。可以根据内存、CPU 利用率或请求速率来扩展集群。在本示例中，我们将配置扩展策略以使用 CPU 利用率。即如果 CPU 利用率达到 20%的平均水平，则自动扩展组将创建更多实例。据此，创建一个名为 config.json 的文件，具体如下：

```
{
    "TargetValue": 20.0,
    "PredefinedMetricSpecification":
    {
        "PredefinedMetricType":
        "ASGAverageCPUUtilization"
    }
}
```

（7）将扩展策略附加到自动扩展组中，具体如下：

```
$ aws autoscaling put-scaling-policy --policy-name cpu20 --auto-scaling-group-name message-service-asg --policy-type TargetTrackingScaling --target-tracking-configuration file://config.json
```

现在，自动扩展组已被配置为当 CPU 利用率平均值超过 20%时即增长。启动配置还可以包括引导步骤，用于安装和配置服务（通常使用某种配置管理工具，如 Chef 或 Puppet），或者将其配置为从私有 Docker 存储库中提取 Docker 镜像。

第 9 章　部署微服务

本章包含以下操作秘笈。
- 配置服务以在容器中运行。
- 使用 Docker Compose 运行多容器应用程序。
- 在 Kubernetes 上部署服务。
- 使用金丝雀部署方式测试版本。

9.1　导　　语

近几年来，我们向用户交付软件的方式发生了巨大变化。仅仅在若干年前，常见的做法还是通过在一组服务器上运行 shell 脚本来部署到生产环境中，这些服务器从某种源代码控制存储库中提取更新。这种方法的问题很明显——很难进行扩展，引导服务器容易出错，并且部署很容易陷入我们不希望的状态，从而给用户带来不可预测的体验。

配置管理系统（如 Chef 或 Puppet）的出现在某种程度上改善了这种情况。现在不必在远程服务器上运行自定义 bash 脚本或命令，而是将远程服务器标记为某种类型的角色，指示它们如何配置和安装软件。自动配置的声明式样式更适合于大规模软件部署。还有一些服务器自动化工具（如 Fabric 或 Capistrano）也已被广泛接受，它们试图将代码推送到生产中的过程自动化，并且对于今天不在容器中运行的应用程序，它们也非常流行。

容器彻底改变了我们交付软件的方式。容器使开发人员可以将其代码与所有依赖项打包在一起，包括库、运行时、OS 工具和配置。这样就可以在无须配置主机服务器的情况下交付代码，从而减少了移动部件的数量，大大简化了过程。

通过容器提交服务的过程被称为不可变基础架构（Immutable Infrastructure），因为一旦建立了镜像，通常就不会更改它。取而代之的是，新版本的服务会生成新的镜像。

软件部署方式的另一个重大变化是十二因素方法论的普及。有关该方法论的详细说明，可访问以下网址：

https://12factor.net/

十二因素（Twelve-Factor，常称为 12F）最初是一组由 Heroku 工程师编写的准则。十二因素应用程序的核心设计为与环境松散耦合，从而产生可与各种日志记录工具、配置系统、程序包管理系统和源代码控制系统一起使用的服务。可以说，十二因素应用程序最普遍被接受的概念是，配置可通过环境变量访问，并且日志可输出到标准输出中。

所谓的"十二因素"具体包括以下概念。

- 代码库。
- 依赖。
- 配置。
- 支持服务。
- 构建、发布和运行。
- 处理进程。
- 端口绑定。
- 并发。
- 可处理性。
- 开发/生产相似性。
- 日志。
- 管理例程。

由此可见，到目前为止，本书已经遵循了十二因素中描述的许多概念。

在本章中，我们将讨论容器、业务流程和调度，以及将更改安全地传送给用户的各种方法。这是一个非常活跃的主题，目前开发人员仍在尝试和讨论各种新技术，但是本章中的秘笈应作为一个很好的起点，特别是如果你习惯于在虚拟机或裸机服务器上部署单体架构应用程序。

9.2 配置服务以在容器中运行

众所周知，服务由源代码和配置组成。例如，用 Java 编写的服务可以打包为 Java 存档（Java Archive，JAR）文件，其中包含以 Java 字节码编译的类文件以及诸如配置和属性文件之类的资源。打包后，即可在运行 Java 虚拟机（Java Virtual Machine，JVM）的任何计算机上执行 JAR 文件。

为了使这种机制能够正常工作，我们必须在运行服务的计算机上安装 Java 虚拟机（JVM）。一般来说，它必须是 JVM 的特定版本。此外，计算机可能还需要安装一些其他实用程序，或者可能需要访问共享文件系统。尽管这些本身不是服务的一部分，但它们确实构成了服务的运行时环境。

9.2.1 理论阐释

Linux 容器是一项技术，允许开发人员使用其完整的运行时环境打包应用程序或服务。容器可以将特定应用程序的运行时与运行容器的主机的运行时分开。

这使应用程序更具可移植性，使服务从一种环境迁移到另一种环境变得更加容易。工程师可以在笔记本电脑上运行服务，然后将其移动到预生产环境中，然后再进入生产环境中，而无须更改容器本身。容器还使你可以轻松地在同一台机器上运行多个服务，因此在应用程序架构的部署方式上提供了更大的灵活性，并提供了运营成本优化的机会。

Docker 是容器运行时和工具集，可让你为服务创建独立的执行环境。当前，还有其他流行的容器运行时被广泛使用，但是 Docker 旨在使容器具有可移植性和灵活性，使其成为构建服务容器的理想选择。

在本秘笈中，我们将使用 Docker 创建一个打包 message-service 服务的镜像。我们将创建一个 Dockerfile 文件并使用 Docker 命令行实用程序来创建一个镜像，然后将该镜像作为容器运行。

9.2.2 实战操作

本秘笈需要执行以下操作。

（1）打开前面章节中的 message-service 服务项目。在项目的根目录中创建一个名为 Dockerfile 的新文件，具体如下：

```
FROM openjdk:8-jdk-alpine
VOLUME /tmp
EXPOSE 8082
ARG JAR_FILE=build/libs/message-service-1.0-SNAPSHOT.jar
ADD ${JAR_FILE} app.jar
ENTRYPOINT ["java","-Djava.security.egd=file:/dev/./urandom",
"- jar","/app.jar"]
```

（2）Dockerfile 文件定义了用于构建 message-service 服务镜像的基础镜像。在本示例中，我们将使用 OpenJDK 8 在 Alpine Linux 镜像的基础上创建镜像。接下来我们将公开服务绑定的端口，并定义在打包为 JAR 文件后运行服务的方式。现在，我们将准备使用 Dockerfile 文件构建镜像。这是通过以下命令完成的：

```
$ docker build . -t message-service:0.1.1
```

（3）可以通过运行 docker images 并查看列出的命令来验证步骤（2）中的命令是否

有效。现在，我们准备通过在容器中执行服务来运行 message-service 服务。这是通过 docker run 命令完成的。我们还将为其提供端口映射，并指定要用于运行服务的镜像：

```
$ docker run -p 0.0.0.0:8082:8082 message-service:0.1.1
```

9.3 使用 Docker Compose 运行多容器应用程序

服务很少单独运行。微服务通常连接到某种类型的数据存储，并且可能具有其他运行时相关性。为了使用微服务，必须在开发人员的机器上以本地方式运行它。要求工程师手动安装和管理服务的所有运行时依赖关系，以在微服务上工作将是不切实际且耗时的。相反，我们需要一种自动管理运行时服务依赖项的方法。

9.3.1 理论阐释

容器使服务更易于移植，因为它可以将运行时环境和配置与应用程序代码作为可交付的工件打包在一起。为了使通过容器进行本地开发的利益最大化，有必要能够声明所有依赖项并在单独的容器中运行它们。这就是 Docker Compose 的目的。

Docker Compose 使用声明性 YAML 配置文件来确定应如何在多个容器中执行应用程序。这将使得快速启动服务、数据库以及该服务的其他运行时依赖关系变得很容易，并且也使得本地开发特别容易。

在本秘笈中，我们将按照 9.2 节的秘笈中的一些步骤为 authentication-service 服务项目创建 Dockerfile 文件。然后，我们将创建一个 Docker Compose 文件，该文件将 MySQL 指定为 authentication-service 服务的依赖项。最后，我们将研究如何配置项目并在一个运行应用程序的容器和另一个运行数据库服务器的容器中以本地方式运行该项目。

9.3.2 实战操作

本秘笈需要执行以下操作。

（1）打开 authentication-service 服务项目，并创建一个名为 Dockerfile 的新文件，具体如下：

```
FROM openjdk:8-jdk-alpine
VOLUME /tmp
EXPOSE 8082
ARG JAR_FILE=build/libs/authentication-service-1.0-SNAPSHOT.jar
```

```
ADD ${JAR_FILE} app.jar
ENTRYPOINT ["java","-Djava.security.egd=file:/dev/./urandom",
"- jar","/app.jar"]
```

（2）Docker Compose 使用一个名为 docker-compose.yml 的文件来声明如何运行容器化应用程序，具体如下：

```
version: '3'
services:
    authentication:
        build: .
        ports:
            - "8081:8081"
        links:
            - docker-mysql
        environment:
            DATABASE_HOST: 'docker-mysql'
            DATABASE_USER: 'root'
            DATABASE_PASSWORD: 'root'
            DATABASE_NAME: 'user_credentials'
            DATABASE_PORT: 3306
    docker-mysql:
        ports:
            - "3306:3306"
        image: mysql
        restart: always
        environment:
            MYSQL_ROOT_PASSWORD: 'root'
            MYSQL_DATABASE: 'user_credentials'
            MYSQL_ROOT_HOST: '%'
```

（3）由于要连接到在 docker-mysql 容器中运行的 MySQL 服务器上，因此我们需要修改 authentication-service 服务配置，以在连接到 MySQL 时使用该主机，具体如下：

```
server:
    port: 8081

spring:
    jpa.hibernate.ddl-auto: create
    datasource.url: jdbc:mysql://docker-mysql:3306/user_credentials
    datasource.username: root
    datasource.password: root
```

```
hibernate.dialect: org.hibernate.dialect.MySQLInnoDBDialect

secretKey: supers3cr3t
```

（4）现在可以使用以下命令运行 authentication-service 服务和 MySQL：

```
$ docker-compose up
Starting authentication-service_docker-mysql_1 ...
```

（5）至此，authentication-service 服务现在应该在容器中以本地方式运行。

9.4 在 Kubernetes 上部署服务

容器允许你将代码、依赖项和运行时环境打包在一个工件中，从而使服务具有可移植性。部署容器通常比部署不在容器中运行的应用程序更容易。主机不需要具有任何特殊的配置或状态。它只需要能够执行容器运行时。在管理生产环境时，将一个或多个容器部署在单个主机上的能力提出了另一项挑战——调度和编排容器以在特定主机上运行并管理扩展。

9.4.1 理论阐释

Kubernetes 是一个开源容器编排工具。它负责调度、管理和扩展你的容器化应用程序。使用 Kubernetes，你不必担心将容器部署到一个或多个特定主机上。取而代之的是，你只需要声明容器需要的资源，即可让 Kubernetes 决定如何进行工作（例如，容器在什么主机上运行、与之一起运行的服务等）。

Kubernetes 源自 Google 工程师发布的 Borg 论文，有关详细信息可访问以下网址：

https://research.google.com/pubs/pub43438.html

上述论文描述了他们如何使用 Borg 集群管理器管理 Google 数据中心中的服务。

Kubernetes 于 2014 年由 Google 作为一个开源项目启动，并且已经被在容器中部署代码的组织广泛接受和采用。

安装和管理 Kubernetes 集群超出了本书的讨论范围。幸运的是，有一个名为 Minikube 的项目使你可以轻松地在开发机器上运行单节点 Kubernetes 集群。即使集群只有一个节点，部署服务时与 Kubernetes 进行交互的方式通常也是相同的，因此任何 Kubernetes 集群都可以遵循本节操作步骤。

在本秘笈中，我们将安装 Minikube，启动单节点 Kubernetes 集群，并部署在前几章中使用过的 message-service 命令。此外，我们还将使用 Kubernetes CLI 工具（kubectl）与 Minikube 进行交互。

9.4.2 实战操作

本秘笈需要执行以下操作。

（1）为了演示如何将服务部署到 Kubernetes 集群中，我们将使用一个名为 minikube 的工具。minikube 工具使得在虚拟机（Virtual Machine，VM）上运行单节点 Kubernetes 集群变得更容易。虚拟机可在笔记本电脑或开发机器上运行。据此，安装 minikube。若在 macOS X 上，则可以使用 HomeBrew 来执行以下操作：

```
$ brew install minikube
```

（2）在此秘笈中还将使用 Kubernetes CLI 工具，因此需要先安装它们。若在 macOS X 上，则可以使用 HomeBrew 来执行以下操作：

```
$ brew install kubernetes-cli
```

（3）现在我们已经为启动单节点 Kubernetes 集群做好了准备。可以通过运行 minikube start 来启动它，具体如下：

```
$ minikube start
Starting local Kubernetes v1.10.0 cluster...
Starting VM...
Getting VM IP address...
Moving files into cluster...
Setting up certs...
Connecting to cluster...
Setting up kubeconfig...
Starting cluster components...
Kubectl is now configured to use the cluster.
Loading cached images from config file
```

（4）将 minikube 集群设置为 kubectl CLI 工具的默认配置。对应的命令如下：

```
$ kubectl config use-context minikube
Switched to context "minikube".
```

（5）通过运行 cluster-info 命令，验证所有配置均正确。对应的命令如下：

```
$ kubectl cluster-info
Kubernetes master is running at https://192.168.99.100:8443
KubeDNS is running at
```

```
https://192.168.99.100:8443/api/v1/namespaces/kube-system/services/
kube-dns:dns/proxy
```

要进一步调试和诊断集群问题,可以使用 kubectl cluster-info dump。

(6) 现在应该能够在浏览器中启动 Kubernetes 仪表板,具体如下:

```
$ minikube dashboard
Waiting, endpoint for service is not ready yet...
Opening kubernetes dashboard in default browser...
```

(7) minikube 工具可使用许多环境变量来配置 CLI 客户端。例如,可使用以下命令评估环境变量:

```
$ eval $(minikube docker-env)
```

(8) 我们将使用在 9.3 节的秘笈中创建的 Dockerfile 文件为我们的服务构建 Docker 镜像,命令如下:

```
$ docker build -t message-service:0.1.1
```

(9) 在 Kubernetes 集群上运行 message-service 命令,告诉 kubectl 要使用的正确镜像以及要公开的端口,具体如下:

```
$ kubectl run message-service --image=message-service:0.1.1 --
port=8082 --image-pull-policy=Never
```

(10) 可以通过列出集群上的 pods 来验证 message-service 命令是否在 Kubernetes 集群中运行,具体如下:

```
$ kubectl get pods
NAME READY STATUS RESTARTS AGE
message-service-87d85dd58-svzmj 1/1 Running 0 3s
```

(11) 为了访问 message-service 命令,需要将其公开为新服务,具体如下:

```
$ kubectl expose deployment message-service --type=LoadBalancer
service/message-service exposed
```

(12) 可以通过在 Kubernetes 服务上列出服务来验证先前的命令,具体如下:

```
$ kubectl get services

NAME TYPE CLUSTER-IP EXTERNAL-IP PORT(S) AGE
kubernetes ClusterIP 10.96.0.1 <none> 443/TCP 59d
message-service LoadBalancer 10.105.73.177 <pending> 8082:30382/TCP 4s
```

（13）minikube 工具有一个方便的命令，用于访问在 Kubernetes 集群上运行的服务。运行以下命令将列出运行 message-service 命令的 URL：

```
$ minikube service list message-service
|-------------|--------------------|------------------------------|
| NAMESPACE   | NAME               | URL                          |
|-------------|--------------------|------------------------------|
| default     | kubernetes         | No node port                 |
| default     | message-service    | http://192.168.99.100:30382  |
| kube-system | kube-dns           | No node port                 |
| kube-system | kubernetes-dashboard | http://192.168.99.100:30000 |
|-------------|--------------------|------------------------------|
```

（14）使用 curl 尝试对服务进行请求以验证其是否正常运行。

恭喜你！你已经在 Kubernetes 上部署了 message-service 命令。

9.5 使用金丝雀部署方式测试版本

多年来，最佳部署实践的改进极大地提高了部署的稳定性。自动化可重复的步骤、标准化应用程序与运行时环境交互的方式，以及将应用程序代码与运行时环境打包在一起，这些措施都使部署比以往更安全、更轻松。

9.5.1 理论阐释

当然，将新代码引入生产环境中并非没有风险。本章讨论的所有技术都有助于防止可预见的错误，但是它们并不能防止实际的软件错误对用户产生的负面影响。金丝雀（Canary）部署是一种减少这种风险，并提高对部署到生产环境中的新代码的信任的技术。

使用金丝雀部署时，首先需要将代码交付给较小百分比的生产流量。然后，你可以监视度量指标、日志、跟踪或其他允许你观察软件工作方式的工具。一旦你确信一切都会按计划进行，就可以逐渐增加接收更新版本的流量的百分比，直到所有生产流量都由服务的最新版本提供。

金丝雀部署（Canary Deployment）一词来自煤矿工人用来保护自己免受一氧化碳或

甲烷中毒的一种技术。以前，矿工在下矿洞之前，会先将一只金丝雀放入矿洞中，进而通过金丝雀能否存活来判断洞中是否含有毒气体。如果有毒气体杀死了金丝雀，则可为矿工提供一个预警信号，表明他们应该撤离。同样，金丝雀部署使我们能够使一部分用户面临风险，而不会影响其余的生产环境。值得庆幸的是，将代码部署到生产环境中时，不会对动物造成伤害。

金丝雀部署过去很难正确实现。以这种方式发布软件的团队通常必须提出某种功能切换解决方案，以将请求发送给正在部署的特定版本的应用程序。幸运的是，容器使金丝雀部署变得容易得多，而 Kubernetes 则使其变得更加容易。

在本秘笈中，我们将使用金丝雀部署方法将更新部署到 message-service 服务应用程序中。由于 Kubernetes 能够从 Docker 容器注册表中提取镜像，因此我们将在本地运行注册表。一般来说，可以使用自托管注册表或诸如 Docker Hub 或 Google Container Registry 之类的服务。首先，我们将确保在 minikube 中具有运行 message-service 命令的稳定版本，然后引入更新并将其逐步推广到 100%的流量。

9.5.2 实战操作

请执行以下操作来设置金丝雀部署。

（1）打开在前面的秘笈中已经处理过的 message-service 服务项目。将以下 Dockerfile 文件添加到项目的根目录中：

```
FROM openjdk:8-jdk-alpine
VOLUME /tmp
EXPOSE 8082
ARG JAR_FILE=build/libs/message-service-1.0-SNAPSHOT.jar
ADD ${JAR_FILE} app.jar
ENTRYPOINT ["java","-Djava.security.egd=file:/dev/./urandom",
"- jar","/app.jar"]
```

（2）为了让 Kubernetes 知道服务是否正在运行，我们需要添加一个活动性探针端点。打开 MessageController.java 文件，并在/ping 路径中添加一种方法来响应 GET 请求，具体如下：

```
package com.packtpub.microservices.ch09.message.controllers;

import com.packtpub.microservices.ch09.message.MessageRepository;
import com.packtpub.microservices.ch09.message.clients.SocialGraphClient;
import com.packtpub.microservices.ch09.message.exceptions.
MessageNotFoundException;
```

```java
import com.packtpub.microservices.ch09.message.exceptions.
MessageSendForbiddenException;
import com.packtpub.microservices.ch09.message.models.Message;
import com.packtpub.microservices.ch09.message.models.UserFriendships;
import org.springframework.beans.factory.annotation.Autowired;
import org.springframework.http.ResponseEntity;
import org.springframework.scheduling.annotation.Async;
import org.springframework.web.bind.annotation.*;
import org.springframework.web.client.RestTemplate;
import org.springframework.web.servlet.support.
ServletUriComponentsBuilder;

import java.net.URI;
import java.util.List;
import java.util.concurrent.CompletableFuture;

@RestController
public class MessageController {

    @Autowired
    private MessageRepository messagesStore;

    @Autowired
    private SocialGraphClient socialGraphClient;

    @RequestMapping(path = "/{id}", method = RequestMethod.GET,
produces = "application/json")
    public Message get(@PathVariable("id") String id) throws
MessageNotFoundException {
        return messagesStore.get(id);
    }

    @RequestMapping(path = "/ping", method = RequestMethod.GET)
    public String readinessProbe() {
        return "ok";
    }

    @RequestMapping(path = "/", method = RequestMethod.POST,
produces = "application/json")
    public ResponseEntity<Message> send(@RequestBody Message
message) throws MessageSendForbiddenException {
        List<String> friendships =
```

```java
socialGraphClient.getFriendships(message.getSender());

        if (!friendships.contains(message.getRecipient())) {
            throw new MessageSendForbiddenException("Must be friends to send message");
        }

        Message saved = messagesStore.save(message);
        URI location = ServletUriComponentsBuilder
                .fromCurrentRequest().path("/{id}")
                .buildAndExpand(saved.getId()).toUri();
        return ResponseEntity.created(location).build();
    }

    @RequestMapping(path = "/user/{userId}", method = RequestMethod.GET, produces = "application/json")
    public ResponseEntity<List<Message>> getByUser(@PathVariable("userId") String userId) throws MessageNotFoundException    {
        List<Message> inbox = messagesStore.getByUser(userId);
        if (inbox.isEmpty()) {
            throw new MessageNotFoundException("No messages found for user: " + userId);
        }
        return ResponseEntity.ok(inbox);
    }

    @Async
    public CompletableFuture<Boolean> isFollowing(String fromUser, String toUser) {
        String url = String.format("http://localhost:4567/followings?user=%s&filter=%s",
                fromUser, toUser);

        RestTemplate template = new RestTemplate();
        UserFriendships followings = template.getForObject(url, UserFriendships.class);

        return CompletableFuture.completedFuture(
                followings.getFriendships().isEmpty()
        );
    }
}
```

（3）现在可以在端口 5000 上启动容器注册表，命令如下：

```
$ docker run -d -p 5000:5000 --restart=always --name registry
registry:2
```

（4）由于我们使用的本地存储库未配置有效的 SSL 证书，因此可以从不安全的存储库中拉出 minikube 来启动，命令如下：

```
$ minikube start --insecure-registry 127.0.0.1
```

（5）构建 message-service 服务 Docker 镜像，然后使用以下命令将该镜像推送到本地容器注册表中：

```
$ docker build . -t message-service:0.1.1
...
$ docker tag message-service:0.1.1 localhost:5000/message-service
...
$ docker push localhost:5000/message-service
```

（6）Kubernetes 部署对象描述了 Pod 和 ReplicaSet 的所需状态。在我们的部署中，将指定希望 message-service 服务 Pod 始终运行 3 个副本，并指定在前面几个步骤中创建的活动性探针。要为 message-service 服务创建部署，可创建一个名为 Deployment.yaml 的文件，其中包含以下内容：

```yaml
apiVersion: extensions/v1beta1
kind: Deployment
metadata:
    name: message-service
spec:
    replicas: 3
    template:
        metadata:
            labels:
                app: "message-service"
                track: "stable"
        spec:
            containers:
                - name: "message-service"
                  image: "localhost:5000/message-service"
                  imagePullPolicy: IfNotPresent
                  ports:
                    - containerPort: 8082
                  livenessProbe:
```

```
            httpGet:
                path: /ping
                port: 8082
                scheme: HTTP
            initialDelaySeconds: 10
            periodSeconds: 30
            timeoutSeconds: 1
```

（7）使用 kubectl，我们将创建以下部署对象：

```
$ kubectl create -f deployment.yaml
```

（8）现在可以通过运行 kubectl get pods 来验证我们的部署是否处于活动状态以及 Kubernetes 是否正在创建 Pod 副本，命令如下：

```
$ kubectl get pods
```

（9）现在我们的应用程序正在 Kubernetes 中运行，下一步是创建一个更新并将其推广到 Pod 的子集。首先，需要创建一个新的 Docker 镜像；在本示例中，我们将其称为版本 0.1.2 并将其推送到本地存储库中：

```
$ docker build . -t message-service:0.1.2
...
$ docker tag message-service:0.1.2 localhost:5000/message-service
$ docker push localhost:5000/message-service
```

（10）现在可以配置一个部署以运行最新版本的镜像，然后再将其推广到其他 Pod 上。